AF443150

New Perspectives in Chemistry and Biochemistry

NEW PERSPECTIVES IN CHEMISTRY AND BIOCHEMISTRY

G. E. ZAIKOV

EDITOR

Nova Science Publishers, Inc.

New York

Senior Editors: Susan Boriotti and Donna Dennis
Coordinating Editors: Tatiana Shohov and Jennifer Vogt
Office Manager: Annette Hellinger
Graphics: Wanda Serrano
Editorial Production: Matthew Kozlowski, Jonathan Rose and Maya Columbus
Circulation: Ave Maria Gonzalez, Vera Popovic and Vladimir Klestov
Communications and Acquisitions: Serge P. Shohov
Marketing: Cathy DeGregory

Library of Congress Cataloging-in-Publication Data
Available upon request.
ISBN 1-59033-449-3

Copyright © 2002 by Nova Science Publishers, Inc.
    400 Oser Ave, Suite 1600
    Hauppauge, New York 11788-3619
    Tele. 631-231-7269   Fax 631-231-8175
    e-mail: Novascience@earthlink.net
    Web Site: http://www.novapubishers.com

Printed in the United States of America

CONTENTS

Institutes Where Research was Done

Institute	Number of page in the volume
N.M.Emanuel Institute of Biochemical Physics Rus.Ac.Sci., Moscow	1; 11; 25; 33; 41; 49; 55; 63; 81; 113
N.N.Semenov Institute of Chemical Physics, Moscow	5; 103
Moscow State University of Applied Biotechnology, Moscow	5
Bashkirian State University, Ufa	81
Kabardino-Balkarian State University, Nal'chik	25; 33; 41; 49; 55; 63; 81
I.Javakhishvili Tbilisi State University, Tbilisi, Georgia	87; 95

PREFACE

Epigraph
How much nonsense did you send to the world?
Leonardo da Vinci
Artist and Inventor, Italy

That was the question that the creator of the Mona Liza asked the inventors and engineers of his time. If we analyse carefully the discoveries of Leonardo da Vinci, he in fact made lots of discoveries as Archimedes, but, if we look closely, the genius of the Renaissance often made mistakes and technical errors himself (for example he thought that man could fly as a bird).

If a man of genius often made mistakes what can we demand from ordinary scientists? Of course the articles in this volume are writeen by not so ordinary scientists, but written by outstanding researchers of different Institutes of the Russian Academy of Sciences. Yet, even in this case not everything that is presented here will have a future. It is better to look upon this problem from the other view: we hope that the majority of the ideas in this volume will be useful and find a scientifical future like the Research-Development-Production.

SURFACES SPIN LABELING – NEW APPROACH TO STUDYING MACROMOLECULES INTERACTION WITH THE SOLID SURFACE AND PROPERTIES OF BOARDER LAYERS IN POLYMER COMPOSITES

A. L. Kovarski, T. V. Yushkina, and A. P. Tikhonov[*]
Emanuel Institute of Biochemical Physics RAS, 119991 Moscow,
Kosygin Str, 4, Russia
[*]Mendeleev Chemical Technology University, 125047 Moscow,
Miusskaya Sq., 9, Russia

There are a limited number of methods for interface layers in filled polymer materials research. Considering that the interface layers present only a small part of the whole material volume, more selective methods compared to the traditional ones (e.g. dielectric and nuclear magnetic relaxation) are required for their investigation. The methods using molecular sensors, localized over microregions of composition material under investigation, seems more suitable for this aim. The paramagnetic (spin) probes and labels which are stable nitroxide radicals have been widely used in study of macromolecular systems both of chemical and biological nature [1-3]. The number of investigations of macromolecules on the liquid/solid interface of synthetic polymer composites was carried out using spin labels i.e. nitroxides covalent bonded to macromolecules [see for example 4-9]. Nevertheless, the application of this technique to the solid filled composites investigation meets several obstacles. First of all the attachment of labels to macromolecules requires the complicated procedure of covalent binding of nitroxide with different macromolecular groups, including low active ones. And secondly, this approach leads to the modification of macromolecules by strongly polar and bulk nitroxide fragments capable of hydrogen bonds formation.

The recent investigations have shown that attachment of spin label to the surface of a filler is much more promising method. It gives a chance to simplify the labeling procedure and allows to investigate the interaction between unmodified macromolecules of different chemical nature and the solid surface of the filler. In present time two methods of the label fixing on the surface of filler particles are worked out. First is based on covalent bonding of nitroxide to chemical groups on the surface. As an example we regarded the attachment of nitroxide radical to the particles of silica dioxide [10,11]. The surface of silica dioxide bears the considerable amount of silanol groups capable to modification. It is this structural peculiarity of SiO_2 particles that used for surface modification and adhesion improvement. This type of nitroxide attachment carried out firstly in work [12] included two base stages – silanol groups modification by ω-halogenchlorinesilane and bonding of stable nitroxide radical 2,2,6,6-tetramethyl-4-aminopiperidin-1-oxyle to halogen carried groups on the surface. According to this procedure the spin labels with different length of the bonding bridge can be obtained. Chemical structure of the spin label used in work [10] is present below.

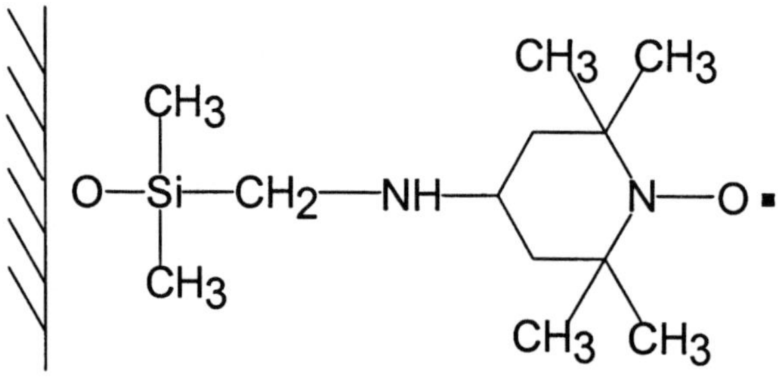

Due to high sensitivity of ESR technique using for obtaining the information about molecular mobility of spin labels, only a small part of silanol groups on the surface can be modified. Therewith the adsorptive characteristics of the surface aren't changing. For example in the works [11] the concentration of spin labels was ~3.10^{17}spin/gr, i.e. one label is situated on the area of about 200 nm^2.

It should be notified that not only the particles of silica dioxide, but also the particles of other mineral substances using as fillers for polymer compositions have OH-groups on the surface capable to chemical modification (e.g. metal oxide). The spin labels can also be attached to the surface of organic polymer particles. As it was reported spin labeled particles of polystyrene latex contained the amino and epoxy groups on the surface have been obtained.

Recently more available approach to spin labeling was suggested. In this approach nitroxide radicals, which can be adsorb strongly on to the surface should be used. As such label the nitroxide radical 3-piperidin-N-methyl-2,2,6,6-tetramethyl-4-oxopiperidin-1-oxile is in use.

This label has charged atoms and is eager to fix on the surface both with hydrogen bonds and ionic forces [10-11]. However these labels can be used only for nonpolar matrixes. It was shown that rotational mobility of surface labels of the both types mentioned is determent by the microreology of liquid or polymer matrix [10-11].

ESR spectra of surface spin labels in air have a peculiarity – strong widening of lines because of interaction between paramagnetic fragment and oxygen.

This widening can be removed by oxygen evacuation as well as by adsorption of macromolecules (and also by adsorption of liquid particles), which isolate the surface from oxygen. Thus, the analysis of ERS spectra gives the answer to the question about forming of the adsorption layer and about the share of surface of filler particles blocked by macromolecules.

There are the examples of problems which can be solved with the aid of surface spin labels:

- The formation of adsorption layer on the surface of solid filler particles under the different conditions of mixtures preparation;
- The analysis of conformation and mobility of macromolecules adsorbed on the surface;

The "layer by layer" analysis of mobility of macromolecules situated at different distances from the surface of filler particles. The determination of number of boarder layers with modified physical properties;

- The investigation of selective adsorption of macromolecules of different structure from the mutual solution.

REFERENCES

1. *Spin Labeling. Theory and Application.* Editor – L.J.Berliner. Academic Press. N.Y. 1976.
2. A.M.Wasserman, A.L.Kovarski, *Spin Labels and Probes in polymer physics chemistry.* M.:Nauka, 1986.
3. Kovarski A.L. Spin probes and Labels. A Quarter of a Century of Application to Polymer Studies. *Polymer Yearbook*, 1996, v.13, pp.113-139.
4. Shimada-S., Horiguchi-K Yamamoto-K. // *Colloid and Polymer Sci.* **276(5)**, 412 (1998).
5. Fox K.K., Robb I. D., Smith R. // *J.Chem. Faraday Trans. Part I,* **70(7)**, 1186 (1974).
6. Robb I. D., Smith R. // *Polymer.***18(5)**, 500 (1977).
7. Clark A.T., Robb I. D., Smith R. // *J.Chem. Faraday Trans. Part I.* **72(6)**, 1489 (1976).
8. Miller W.G., Rudolf W.T., Veksli Z., Coon D.L., Wu C.C., Liang T.M. in *Molecular Motion in Polymers* by ESR / Ed. by Boyer R.F. and Keinath S.E. Chur.: Harwood. MMI Press Symposium Series. V. 1. 1978.
9. Chikhi M., Afif A., Hommel H., Legrand A.P. // *Colloid Polym. Sci.* **277(10)**, 965 (1999).
10. A. L. Kovarsky, T. V. Yushkina, V.V.Kasparov, A. B. Shapiro, and A. P. Tichonov// *Vysokomol. Soedin., Ser.A,* **43(3)**, 264 (2001).
11. T.V.Yushkina, A.L.Kovarsky, V.V.Kasparov, A.B.Shapiro, and A.P.Tichonov // *Vysokomol.Soedin., Ser.A,* **43(11)**, 1184 (2001).
12. Mingalev P.G., Fadeev F.Yu., Staroverov S.M., Lisichkin G.V., Lunina E.V. // *J.Chromatography.* **646**, 267 (1993).

CRYOIMMOBILIZATION OF BIOLOGICALLY ACTIVE COMPOUNDS ON POLYMER CARRIER

E. G. Rozantsev, O. N. Belyatskaya, L. I. Bulatnikova,
L. V. Bykova, T. V. Ivanova
State University of Applied Biotechnology, Moscow, Talalikhin' str. 33

A. L. Iordanskii, and R. Yu. Kosenko
Institute of Chemical Physics of RAS, Moscow, Kosygin' str., 4

SYNOPSIS

For designing of polymer systems with immobilized enzyme, the cryo designed films of polyvinylalcohole (PVA) with bull pepsin or protosubtiline were studied. The films were treated by a freezing-melting cyclic procedure. On chemical and morphological levels the PVA films were analyzed by both FTIR ATR technique and equilibrium water swelling data. In addition, morphological features of PVA were examined by scanning electron microscopy and stress-strain curves were obtained by the "Instron-TM-SM" machine.

Administration of enzymes reduced the network degree of cryo-formed films while the proper cryotreatment led to the enzyme immobilization and the morphological change of polymer matrix. The enzyme activities slightly decreased up to 80-85% but both thermostability and PH range of enzyme operation increased. The influence of cryolisis on stress-strain characteristics of PVA- enzyme films were shown.

INTRODUCTION

A technical use of *enzymes* and other biologically active compounds (BAC) in immobilized form has a number of advantages, already described elsewhere /1 ,2/.

For the most complete realization of these advantages such as prevention of BAC desorption into contacting medium, opportunity of reusing etc, the chemical fixation of BAC on a substrate is relevant. Usually the formation of covalent bonding between polymer and

BAC is affected by 1) high-temperature chemical reactions responcible for BAC inactivation, 2) application of toxical and ecologically nonfreandly solvents, and 3) reaction initiators, that it is undesirable at manufacture of the biomedicine materials or food- industry polymers.

In recent years, the large attention is given to both the use of the low- temperature chemical reactions and the study of features for interactions at above conditions. The investigation of cryochemical processes allows to increase a reaction ability of high molecular substances and to ensure the conditions for the manufacture of products with desirable properties. It is especially important at realization of chemical reactions limiting or excluding application of high temperatures, for example, for immobilization of label BAC.

Polyvinyl alcohol (PVA) is widely used in the capacity of substrates-carriers and matrices for immobilization of various BAC. It is caused by its specific properties - the presence of functional hydroxyl groups, opportunity of controllable water sorption, fat resistance, high mechanical characteristics of films based on PVA.

Recently, authors/4,5/ have shown that at the influence of low temperatures (temperature of phase solvent transition) on water PVA solutions the cryolitical processes are developed in system, as the result the formation of the active paramagnetic centres - free radicals and ions - radicals with strongly located non-pair electrons capable to cooperate with macromolecular links and cryolize components tafes place. The process is accompanied the destruction and structuring of polymer, therefore there is the formation cross links between macromolecules. It was shown, that the rather deep changes in system are observed at one-cyclic influence of low temperature (5 - 7 days) or at short-term (10-60 minutes) cyclic realization of cryolize process. The increase of a number of cycles "freezing - defrostation" allows to intensificate the cryolize process at the increase of the interaction acts in formed active particles among themselves and with entered components; more eficcenly this process is proceeding at the transition from water to ice.

Taking into account above mentioned the aim of this paper is the study of PVA modification in presence of the biological active compounds at low temperatures for the directed regulation of the materials properties.

EXPERIMENTAL PART

The PVA samples have an average MW=60000 and the content of acetatic groups by 1,5 %. Proteolitical enzymes are microbial protosubtilin and the bool pepsin with molecular weight 35.000.

PVA films were received by casting FROM FILTERED 12 %WT WAter PVA solution and PVA compositions with enzymes in quantity 5 %wt (on polymer) on a glass surface. Subsequently prepared samples were treated by cyclical freezing at 255K during 4 cycles, including freesings (60 min) and defrostations (10 min). A control films were formed at 298K. The structurization and net-working of polymer determined from equilibrum swelling in water at 323 K to except molecular interactions in PVA at the cost of hydrogenic bonds. The investigation of overmolecular structures has carried out by electron microscope "Tesla BS-313" at 70000 ^{x}.

Mechanical properties of PVA samples were studied by the universal test machine "Instron TM-SM" at speed of the clamp (slip) movement - 50mm / min and temperature 298K.

Chemical and structural changes in PVA were registered with FTIR spectroscopy as well in a mode of transmission or in a mode of ATR technique. IR-spectra of samples were registered on spectrophotometer with the Fourie transformation ("Bruker IFS48") at the resolution $2cm^{-1}$ and with 250-400 number of scannings. The FTIR ATR spectra were obtained by "Specac F/N 11000" ; the recording has been carried out at a reflection angle equals 45^0 with a nimber of reflections - 25. The band for absorption of CH_2 group with maxima in the range of 2943 and 2910 cm^{-1} IS served as an internal standard at the analysis of samples spectra in adequate conditions.

The samples were extracted by water at 298K during 6 hours for quantitative determination of enzyme connected with polymer PVA film.

The proteolitical activity of enzyme was determined by a standardized Kunitts' method /6/ based on spectrophotometric definition of the quantity of aminoacid (tirozin) formed in result of casein phermentative hydrolysis.

RESULTS AND DISCUSSION

The PVA films formed by freezing are not dissolved in water and limitely swell; the composition of a gel-fraction in the polymer comprises 70-80%, whereas all films generated at 298K completely pass in a solution. At the addition of enzyme in polymer composition in quantity 5 mass.% (on polymer) the cryothreatment of polymer also results to the formation of spatial structure. However the content of a gel-fraction is reduced up to 60-65 % and the equilibrum swelling degree grows with 480% up to 520% and 580% for the samples with pepsin and protosubtilin respectively (Fig. 1).

The reduction of the network degree occurs, apparently, by virtue of the enzyme interaction with active polymer centers (cryolizates) formed during an influence of low temperatures on the polymer system, the reduction by virtue of these acts of polymer-polymer interaction and, in a result, the decrease of network density.

On a molecular level, the polymer transformations which proceed during the cryolitic action on system have been proved at the analysis of IR-spectra of PVA films.

The comparison of IR-spectra of PVA films formed at room temperature (RT) and after cryotreatment (CT) has shown (Fig.2) that in a result of influence of low temperature in PVA there is the band of absorption at 1660 cm^{-1}, which is characteristical for carbonyl groups, belonging probably to the ester bonds (Fig.2a). Their relative increase in a film surface has made 29 % and in volume - 62 % on data of FTIR ATR.

Besides, during the cryothreatment the appreciable charge of hydroxyl groups is observed, that it was registered by the reduction of the area of hydroxyl peak (3340 cm^{-1}) (Fig.2b). From FTIR ATR data the relative areas have made 6,70 and 6,17 for RT-and CT-films respectively, i.e. the hydroxyl-group reduction in a result of system cryothreatment makes - 8%. The charge of OH-groups is apparently connected to their interaction as with active cryolizates as with reaction between two hydroxyle groups of PVA macromolecules resulting in the ester-link formation.

The analysis of a differencial spectrum of enzyme-containing films formed at the low (< O C°) temperatures and control PVA films shows the occurrence of bands of middle and weak intensity at 1656 and 1539 cm^{-1} which are characteristic for amid-1 and amid-2 of the protein bonds /7,8/ (Fig.3). At both ways of a film formation the addition of enzyme in polymer brings for the small increase of intensity in the range of frequencies between 3650 and 3000 cm^{-1}, where alongside with oscillation of OH-groups (PVA) described by a band with a maximum 3340 cm^{-1} the bond oscillation of H-bonded NH-groups are realized with a absorption maximum at 3300 cm^{-1} /3/.

The analysis of IR-spectra in the range of 2920, 1420, and also at 1660 and 1670 cm^{-1} is of interest for discussion of the mechanism of enzyme immobilization process in a PVA matrix. The bands 2920 and 1420 cm^{-1} belong to the oscillation of CH_2 - groups. The change of their intensity in differencial spectrum of cryotreated samples containing enzymes can be related with the effect of radical reactions in the system at the low temperatures and the enzyme interaction with formed free radicals in polymer. The interaction of lizin protein group with residual acetate groups of PVA is possible also because the reduction of intensities at $1660-1670 cm^{-1}$ (belonging to aceto-groups) was testified.

The qualitative spectra analysis of proteins, added in a PVA films at room and negative (below zero) temperatures shows the absence of displacement from 1650 (amid-1) to 1630 and from 1540 (amid-2) to 1520 cm^{-1} characteristic for β-protein structure / /. It allows to consider, that the conformational state of the enzyme in PVA mainly corresponds to the α-spiral form.

The enzyme influence on the secondary structure of PVA was estimated by an account of the syndiotacticity. The syndiotacticity index - Sd - is the percentage of the sindiotactic diads was expected under the formula /10/

$$Sd = 72,4 \left(A_{916} / A_{860} \right)^{0,43}$$

A_{916} and A_{860} - integrated intensity of bamds at 916 and 860 cm^{-1} respectively.

These results testify that the formation of films at low temperatures does not practically influence Sd quantity. It indicates that the native structure of protein in PVA is formed so that to reduce to a minimum of the structural changes into macromolecular spaceregularity.

At the same time, the criolytic influence on system causes the changes in the morphology of supermolecular formations. As it is visible from electron microscopic data (Fig. 4) the fibrilic structure is predominantly formed (Fig. 4a) in films received at 298 K, while at the freezing the system differs by mesh structural formations consisting fron anyzodiametric elements with disordered relative position (Fig. 4b). The decrease of the order degree in PVA macromolecules is promoted both by the fact of the structure formation at low temperature when the mobility of polymer circuits are hindered and by the formation of cross links.

Administration of enzyme into system also effects the morphology of supermolecular polymer structure during the film formation at low temperatures. Alongside with the fibrilic formations that is tipical for PVA, the globular supermolecular structures are observed (Fig. 4в).

The change of the polymer structure occurring during the cryolize renders the significant effect to a mechanical properties of the films. Thus the character of deformation curve is practically identical for the films, formed in different temperature modes - curves "pressure - deformation" look like, characteristic for glass polymer state with a brightly expressed site of compelled elasticity. The significant distinctions are observed for the quantities of tensile strength and relative film elongation : the films formated at low temperatures have the reduced resistant characteristics, which are mostly displayed at the enzyme introduction in PVA-composition (curve 3,4 Fig. 5). The deformation parameters are a little bit increased for the samples with pepsin. At administration of protosubtilin deformation grows with 25 % up to 100 % that is apparently connected with plastisizing effect of enzyme which is displayed for protosubtilin in a greater degree.

The films produced by cryolize of PVA-enzyme compositions have biological activity. The immobilization degree estimated by the determination of enzyme activity makes 80-85 %. By special experiense, it was shown that an investigated temperature of the film cryoformation does not render to activity of free enzyme. The investigation of hydrolysis kinetics for a number of proteins (albumin, fibrine, casein and trim textile (or fabric) at the immobilized enzyme presence has shown that the hydrolysis is rather effectively realized but with the smaller speed than in case of free enzyme. It is apparently connected to the smaller speed of the substrates diffusion to active centres of the immobilized enzymes, being in polymer mass as against the molecules of free enzyme directly contacting with substrate. It is also connected with partial screening of active centres by PVA macromolecules.

The study of the immobilized enzymes activity shows that the enzyme thermostability is increased in comparison with free enzyme and pH-range of continua in which enzyme had haven high catalizing action is extended. With increase in the incubation temperature the immobilized enzyme slowly grows in the begining, and since 40^0C an sharp increase of its activity is observed as well as the maximal activity of the immobilized enzyme is observed at 60^0C when the activity of native enzyme appreciably falls (Fig.6).

SUMMARY

Thus, the investigation of low temperatures effect (the temperatures are lower than ones of the phase solvent transition) on system PVA - enzyme - water has allowed to establish the laws of structure change and the polymer properties in criolize and to produce film materials having biological activity. It has been shown that under the low temperatures effect the formation of cross links between macromolecules and an enzyme immobilization by polymer matrix take place. It allows to develop the enzimatic systems with the improved service characteristics.

REFERENCES

1.   V.V.Korshak, M.I.Shtilman. *Polimers in immobilization and updating of natural connections*, M.; "Ed. Science", 1984, 262c.

2.   I.V.Berezin and other. *Immobilized enzymes*. M.; Higher Shool, 1987, 160c.

3.   G.B.Sergeev, V.A.Batiuk. *Cryochenistry*. M.; Chemistry, 1978,296c.

4.   L.I.Bulatnikova and other. Macromolecular. Compounds *(Visokomolek. Soed.).*, 1983, т. Б25, 86 c.614/

5.   V.E.Gul' and other. *Visokomolek. Soed.*,1976, т. А18, c.118.

6.   V.N.Orekhovich, *Modern methods in biochemistry*, M.; Medicine, 1969, 119c.
7.   K.Nakanisi, *Infra-red spectra and structure organic of connections*. M. The World. 1965, 209c.
8.   Nenzath H., Bailey K. *Proteins Chemistry,*. Vol 2. NJ Pergamon Press 1953, p.375
9.   C.H.Bamford, A.Etliot, W.E.Hanby Tynthetic *Polypeptides*. 1956 NY-London Pergamon Press
10. K.Nakamae, T. Nishino, H.Onkubo, I.Matsuzawa, K.Jamanza, 1992, № 33, № 12, p.2581.

THE FRACTAL ANALYSIS OF THERMO-OXIDATIVE DEGRADATION PROCESS OF POLYMERIC MELTS

G. V. Kozlov, G. B. Shustov[1), and G. E. Zaikov[2)
1. Kabardino-Balkarian State University,
Nalchik – 360004, Chernishevsky st., 173, KBR, Russia
2. Institute of Biochemical Physics, Russian Academy of Sciences,
Moscow – 119991, Kosygin st., 4, Russia

Abstract

The fractal dimension of a macromolecular coil is used for the characterization of polymeric melt structure. It is shown that the kinetics of the oxygen consumption is defined by both chemical constitution of polymer and its structure. The quantitative analysis of the kinetical curves of the oxygen consumption in the thermooxidative degradation process is carried out in the framework of the fractal approach.

Keywords: Polymeric melt, structure, macromolecular coil, oxygen consumption, thermooxidative degradation, fractal dimension.

INTRODUCTION

As it is well known [1, 2], the analysis of the experimental data on oxygen consumption at the thermooxidative degradation of different polymers both in open, and in closed system has shown that there are three main types (under the other data – four types [3]) of kinetic curves "amount of consumed oxygen-time" (N_{O_2} -t). In the present paper one of the indicated types of curves $N_{O_2}(t)$ will be considered: curves exponentially decreased with time (autodecelerated), which are described by the following equation [1]:

$$N_{O_2} = N_{O_2}^{\infty}\left(1 - e^{-kt}\right), \tag{1}$$

where $N_{O_2}^{\infty}$ is limiting amount of oxygen which is capable to be consumed at the oxidation, k is an effective constant of oxidation rate.

The dependence (1) is usually explained by the heterogeneity of polymer which defines the noticeable differences in oxidation rates of its separate sites. As more reactive sites are oxidized first of all, after their practically complete transformation the rate of the summary process strongly reduces. Besides it was exhibited, that the value $N_{O_2}^{\infty}$, determined with the help of formula (1), can be many times as lower than the concentration of monomer units in the individual polymer: during the experiment some part of these units behaves as an inert substance [2].

The value k depends on the temperature and polymer structure. From the mentioned above follows that in essence both N_{O_2} and k are simply fitting parameters, as their quantitative relation with polymer structure is not clarified. Till the recent time the situation became complicated by the absence of a quantitative structural model of polymers, especially when the consideration is about the structure of their melts. However, the last years, development of the methods of the fractal analysis for the description of polymer structure and properties [4, 5] allows to hope for its successful application in this case. The purpose of the present work is to clarity the structural sence of parameters $N_{O_2}^{\infty}$ and k for a polymeric melt on an example of two block-copolymers polyarylatearylenesulphonoxide (PASSO), obtained by the different methods of polycondensation at three testing temperatures [6].

EXPERIMENTAL

PASSO, obtained by low-temperature (PASSO-1) and high-temperature (PASSO-2) polycondensation have been studied. These block-copolymers are synthesized out of diane, mixture (1:1) dichloroanhydrides of tere- and isophthalic acids, dihydroxil-containing olygoarylenesulfonoxide on the basis of diane and 4,4'-dichlorodiphenylsulfone with a molecular weight 4600 of following constitution:

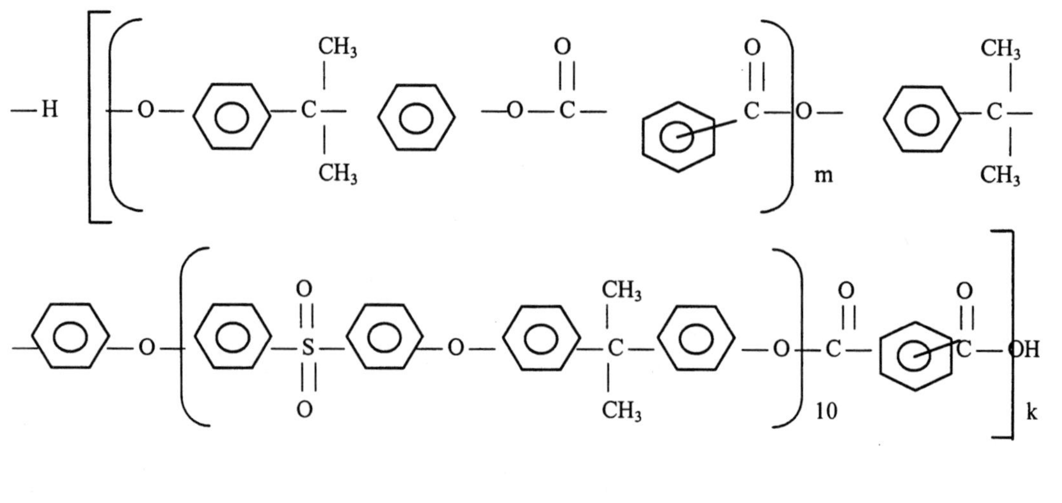

The average molecular weight $\overline{M_w}$ is determined by method of approaching to equilibrium (Archibald method) in ultracentrifuge 3170 of corporation MOM (Hungary). The values $\overline{M_w}$ are equal to 76×10^3 for PAASO-1 and 69×10^3 for PAASO-2 [5]. The glass transition temperature T_g of studied copolymers is defined by the dielectric method. The studies are carried out by quantometer BM-560 "Tesla" at frequency 1 MHz in temperature range 293-573 K [6]. The value T_g is equal to 471 K for PAASO-1 and 491 K – for PAASO-2.

For the studies of the thermooxidative degradation processes in air the ampullary technique is used. The working volume of ampoules is equal to 3×10^{-5} l. The average initial contents of oxygen make up the value 2.5-3.0 mol. O_2/mol of polymer. The kinetic curves of the oxygen consumption $N_{O_2}(t)$ are obtained at temperatures 623 and 723 K [6].

The plotting of logarithmic anamorphoses of kinetic curves $N_{O_2}(t)$ allows to determine the constant of the reaction rate k_d for the thermooxidative degradation.

RESULTS AND DISCUSSION

Let's consider the parameters characterizing a polymeric melt structure. The studies of the thermooxidative degradation process of PAASO is carried out within the temperature range 623-723 K and the temperature range of glass transition of these copolymers is equal to 472-474 K. As it is known [7], the temperature of so-called transition "a liquid 1 – liquid 2" T_{ll} can be estimated as follows:

$$T_{ll} \approx \left(1.20 \pm 0.05\right)T_g. \qquad (2)$$

From equation (2) and quoted above data the condition $T>T_{ll}$ follows. At T_{ll} there occurs a transition of polymeric melt from "liquid with fixed structure" (where the residial structural ordering is observed [7]) to a true liquid state or "unstructured liquid" [8]. Nevertheless, "unstructurity" of melt at $T>T_{ll}$ relates to the absence of supermolecular structure, but the structure of a macromolecular coil in melt remains the important structural factor (in essence, unique at $T>T_{ll}$).

The structure of a macromolecular coil, which is a fractal object [9], may be precisely described with the help of its fractal (Hausdorff) dimension Δ_f, describing the distribution of units of a coil in space [10]. The estimation of the value Δ_f can be made as follows [11]. The formal kinetics of chemical reactions can be described by the following equation:

$$\frac{dQ}{dt} = k_d\left(1-Q\right). \qquad (3)$$

where Q is a reaction degree of conversion, t is its duration, k_d is a constant of the reaction rate.

The general fractal relationship also used for the description of kinetics of the chemical reactions, looks like [5]:

$$Q \sim t^{(3-\Delta_f)/2}.\tag{4}$$

Differentiating relationship (4) by time t and equating the derivative dQ/dt to a similar derivative in equation (3), we shall receive [11]:

$$t^{(\Delta_f-1)/2} = \frac{C_1}{k_d(1-Q)},\tag{5}$$

where C_1 is a constant, which can be estimated from the boundary conditions [11], and the values k_d are accepted under the data of [6].

From the mentioned above description follows, that the thermooxidative degradation process for the considered copolymers proceeds not in an Euclidean, but in a fractal space with dimension Δ_f. In the latter space the process of degradation can be presented by the way of "devil's staircase" [12]. Its horizontal segments correspond to the time intervals, where the reaction does not occur. In this case the degradation process is described with the usage of fractal time t, which belongs to the points of Cantor's set [13]. If the reaction is considered in an Euclidean space, the time belongs to the set of real numbers.

For the description of the evolutional processes with fractal time the mathematics of fractional differentiation and integration will be used [13]. As it is shown in paper [14], in this case the fractional exponent ν coincides the fractal dimension of Cantor's set and indicates a fraction of the system nonchanged all during the time of evolution t. We remind that the Cantor set is considered in one-dimensional Euclidean space ($d=1$) and consequently its fractal dimension $d_f<1$ by virtue of fractal definition [12]. For the fractal objects in Euclidean spaces with higher dimensions ($d>1$) as ν it is necessary to accept the fractional part d_f (in our case Δ_f) or:

$$\nu = \Delta_f - (d-1).\tag{6}$$

Then, value ν characterizes the fraction of a fractal (macromolecular coil), unchanged during the process of degradation. Apparently, the fraction of a macromolecular coil β which is breaking up during the destruction, is determined so:

$$\beta = 1 - \nu = 1 - \left[\Delta_f - (d-1)\right] = d - \Delta_f.\tag{7}$$

or, as in the considered case $d=3$:

$$\beta = 3 - \Delta_f.\tag{8}$$

Further it is naturally to assume that oxygen is consumed in a part of a macromolecular coil which is breaking up during the destruction. Then, having calculated the value $N_{O_2}^T$, which is necessary for the oxidation of all aliphatic and aromatic groups of PASSO (which is equal to 24.1 moles O_2/mole of polymer) it is possible to determine the parameter $N_{O_2}^{\infty}$ as follows:

$$N_{O_2}^{\infty} = \beta N_{O_2}^{T}. \tag{9}$$

The change Δ_f means the variation of the compactness of a macromolecular coil [4]. We should expect that the lowering Δ_f, defining the decrease of the coil compactness, will result in increase N_{O_2} and, therefore, k_d, that allows to replace k_d by the structural parameter. It is possible to characterize the compactness degree of a macromolecular coil with the help of its density ρ. As it is known [12], the value ρ of a fractal is determined as follows:

$$\rho \sim R_g^{\Delta_f - d}, \tag{10}$$

where R_g is a gyration radius of a macromolecular coil. For simplicity further, it was accepted R_g equal to 100 relative units (the values $\overline{M_w}$ for PASSO-1 and PASSO-2 are close) and density ρ was determined from the following equation for both polymers and all testing temperatures:

$$\rho = 8.2 R_g^{\Delta_f - d}, \tag{11}$$

where coefficient 8.2 is defined empirically.

Now the equation (1) with the allowance of the considered geometrical factors can be written so:

$$N_{O_2} = \left(3 - \Delta_f\right) N_{O_2}^{T} \left(1 - e^{-t/\rho}\right). \tag{12}$$

The index of an exponential curve $-(t/\rho)$ is written in such a form because the action ρ is opposite to the action t: the increase ρ reduces the oxygen consumption and on the contrary. In Fig. 1 the experimental dependences $N_{O_2}(t)$ for both studied polymers are given at three temperatures of tests. As it is possible to see all kinetic curves $N_{O_2}(t)$ belong to the considered type: the value N_{O_2} exponentially decreases during the degradation. Such behaviour dN_{O_2}/dt uniquelly indicates the proceeding of the thermooxidative degradation process in a fractal space [15].

In Fig. 2 the generalized kinetic curve of the oxygen consumption for PASSO-1 and PASSO-2 at three temperatures of tests is given. As it is possible to see the experimental results are well described by the equation (12), not containing the empirical coefficients.

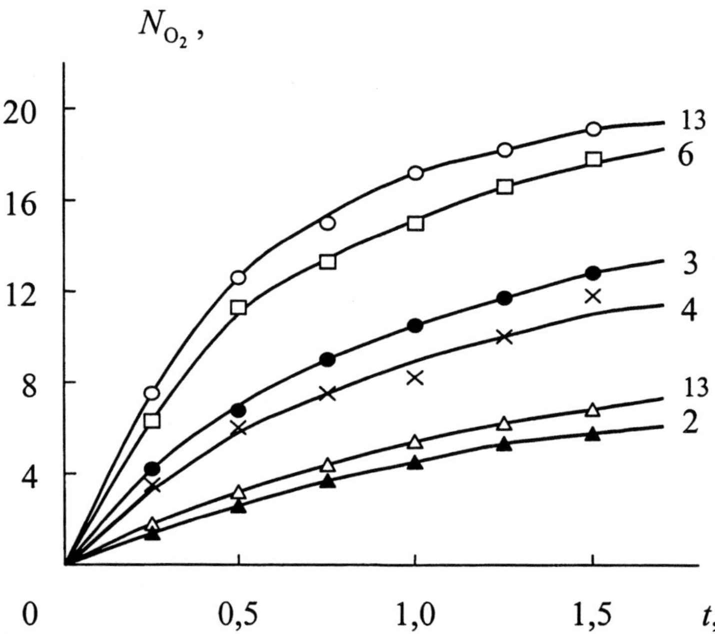

Fig. 1. Dependences of consumed oxygen amount N_{O_2} on time t for PASSO-1 (1,3,5) and PASSO-2 (2,4,6) at the temperatures 623 (1,2), 673 (3,4) and 723 K (5,6).

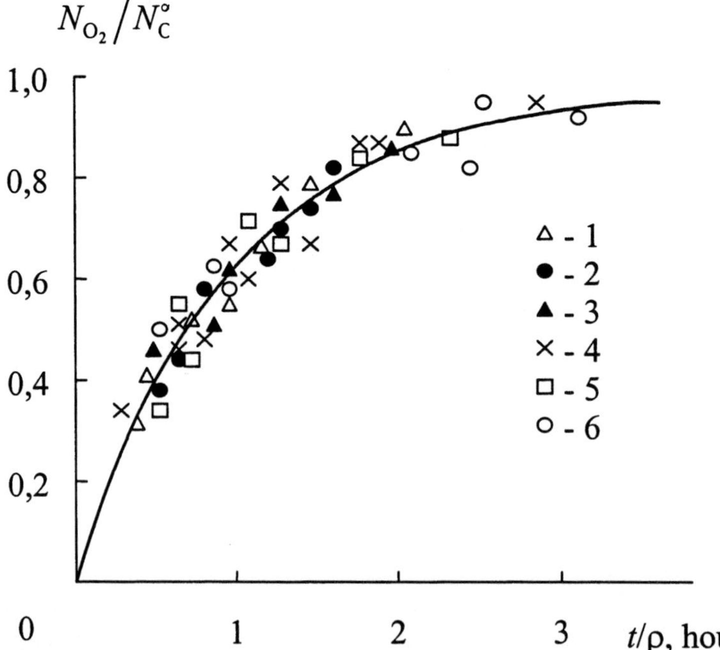

Fig. 2. A generalized kinetic curve of the oxygen consumption corresponding to the equation (12), for PASSO-1 (1-3) and PASSO-2 (4-6) at the temperatures 623 (1,4), 673 (2,5) and 723 K (3,6).

CONCLUSIONS

The results of the present work have shown the necessity of application of the methods of the fractal analysis for the description of the thermooxidative degradation process of polymeric melts. The kinetics of the oxygen consumption is defined by the chemical constitution of polymer $\left(N_{O_2}^T \right)$ and its structure (Δ_f and ρ).

REFERENCES

1. Emanuel N.M. *Vysokomolek. soed.* A, 1985, v. 27, № 7, p. 1347-1363.
2. Shlyapnikov Yu.A., Kirjushkin S.G., Mar'in A.P. *Antioxidative stabilization of polymers* (in Russian). Moscow, Khimiya, 1986, 256 p.
3. Kovarskaya B.M., Blyumenfeld A.B., Levantovskaya I.I. *Thermal stability of heterochain polymers* (in Russian). Moscow, Khimiya, 1977, 276 p.
4. Novikov V.U., Kozlov G.V. *Uspekhi khimii*, 2000, v. 69, № 4, p. 378-399.
5. Novikov V.U., Kozlov G.V. *Uspekhi khimii*, 2000, v. 69, № 6, p. 572-599.
6. Batyrova H.M. *Diss.... kand. khim. nauk.* Moscow, MkhTI, 1985, 160 p.
7. Bershtein V.A., Egorov V.M. *Differential scanning calorymetry in physic-chemistry of polymers* (in Russian). Leningrad, Khimiya, 1990, 256 p.
8. Lobanov A.M., Frenkel S.Ya. *Vysokomolek. soed.* A, 1980, v. 22, № 5, p. 1045-1057.
9. Baranov V.G., Frenkel S.Ya., Brestkin Yu.V. *Dokl. AN SSSR*, 1986, v. 290, № 2, p. 369-372.
10. Vilgis T.A. *Physica A*, 1988, v. 153, № 2, p. 341-354.
11. Ligidov M.H., Kozlov G.V., Bejev A.A. *Izvestiya VUZov – Khimiya*, 2001, v. 44, № 3, p. 27-30.
12. Feder E. *Fractals* (in Russian). Moscow, Mir, 1991, 256 p.
13. Bolotov V.N. *Pisma v ZhTF*, 1995, v. 21, № 10, p. 82-84.
14. Nigmatullin R.R. Teoret. *I matemat. Fiz.*, 1992, v. 90, № 3, p. 354-362.
15. Klimko P.W., Kopelman R. *J. Phys. Chem.*, 1983, v. 87, № 23, p. 4565-4567.

THE THERMO-OXIDATIVE DEGRADATION OF SOLID-STATE POLYMERS: THE STRUCTURAL ANALYSIS WITHIN THE FRAMEWORK OF WITTEN-SANDER MODEL

G. V. Kozlov, V. N. Shogenov, V. V. Afaunov, and G. E. Zaikov
Kabardino-Balkarian State University,
Chernishevsky st., 173, 360004, Nalchik, KBR, Russia

Institute of Biochemical Physics Russian Academy of Sciences,
Kosygin st., 4, 117334, Moscow, Russia

Abstract

The correctness of the application of a fractal model based on the use of Witten-Sander cluster for the identification of amorphous state structure of polymers for the description of the thermo-oxidative degradation in solid-state polymers has been shown. The last process proceeds in a fractal space much more slowly, than in an Euclidean space. Essentially, the represented treatment is a fractal model of the structural stabilization of polymers.

INTRODUCTION

The processes of the thermo-oxidative degradation in solid-state polymers take the important part at thermal aging of these polymers [1]. Such processes result in the reduction of molecular weight and, as a result, to the deterioration of the whole complex of operational properties of polymer products [2]. In the processes of the thermo-oxidative degradation the diffusive processes controlling a carry of an oxidant in a matrix of solid-state polymer, take the important part [1].

As it is known [3], the structure of solid-state polymers is a fractal in an interval of the linear scales from several Ångströms up to several tens of Ångströms. The sizes of free volume holes through which the diffusion of oxidants (especially the gaseous, for example, oxygen) occurs to the reactive sites of polymeric macromolecules, get into this interval. The laws, describing the transport in the fractal objects, essentially differ from the appropriate laws for Euclidean objects [4]. In this connection we shall give two examples. If to consider a trajectory of a molecule of gaseous diffusate as random walk with a root-mean-square step ξ, the number of such steps N_w is determined by relationship [5]:

$$N_w \sim \xi^{d_w},\qquad(1)$$

where d_w is a dimension of random walk.

For Euclidean spaces $d_w=2$ irrespectively from their dimension, whereas for fractals this dimension is much more than 2 [5]. Thus, the number of steps N_w at the identical ξ for fractals will be greater than for an Euclidean object. Or else, in a fractal space the molecule of a diffusate (oxidant) will need to make much more steps for the achievement of a reactive site of polymeric macromolecule, that should lead to the delay of the thermo-oxidative degradation.

The number of sites $\langle S \rangle$, visited by random walk, defines like this [5]:

$$\langle S \rangle ; N_w^{d_s/2},\qquad(2)$$

where d_s is a spectral dimension describing of the object degree connectivity [6].

For Euclidean objects with dimension $d=3$ $d_s=3$, whereas for fractal objects $d_s<2$. From relationship (2) it follows that the number of sites visited by random walk in a fractal space will be smaller than in an Euclidean space. Conformably to a question of the thermo-oxidative degradation it means that in the first case the diffusate molecule can contact to much smaller number of reactive sites in solid-state polymer, than in the second one. Therefore we expect that the rate of the thermo-oxidative degradation for solid-state polymers that have a fractal structure, will be essentially lower than assumed one for the Euclidean solid bodies.

In paper [7] it was shown that the structure of solid-state polymers is simulated by the set of Witten-Sander clusters (clusters WS). The processes of transport and reaction for such clusters are considered in detail in paper [8]. So, the purpose of the present paper is the description of the thermo-oxidative degradation during thermal aging in frameworks of the fractal analysis (cluster WS is a fractal object with dimension ~ 2.5 [5]) by an example of the amorphous glassy and semicryslalline polymers.

EXPERIMENTAL

The semicrystalline industrial gasophase polyethylene of high density (HDPE) of mark 276 (degree of crystallinity ~ 0.69), stabilized by nonchain antioxidant having structure Fe/FeO (Z) with the contents Z 0.01 mass. % (HDPE+Z) is used [9]. Long thermal aging (about 35 days) was carried out in a climatrone with good ventilation of air at 353 K. The melt flow index (MFI) was measured by the automatic capillary viscosimeter IRT-M under a load of 5 kg and at temperature 463 K. According to the data of the firm of "Union Carbide"

(USA) [10] the average weight molecular mass $\overline{M}_w$ for HDPE it is possible to estimate from the following empirical relationship:

$$\lg \overline{M}_w = \lg 129000 - 0,263 \lg MFI .\tag{3}$$

HDPE was a compression moulded into the test specimens measuring 4×6×50 mm for thermal aging.

The polyarylate on the basis of phenolphtaleine, diane and the mixture of dichloranhidrides of iso- and terephtalic acids (PAr) produced by a method of nonequilibrium polycondensation with intrinsic viscosity [η]=1.0 dl/grm is used as the amorphous glassy polymer [11]. Values [η] for initial PAr and the samples subjected to thermal aging during the various intervals of time, are determined for 0.5 % solutions in chloroform at 298 K with the help of the viscosimeter Ubbelode. The samples of PAr for thermal aging tests with the sizes of 4×6×50 mm are produced by molding at temperature ~ 543 K and pressure ~ 100 MPa. Thermal aging was carried out in air at temperature ~ 450 K with the maximum duration of 1000 hours (~ 42 days). The molecular weights of MM of samples of PAr are determined according to the equation of Mark-Houwink [12]:

$$[\eta] = 3,35 \cdot 10^{-4} MM^{0,67} .\tag{4}$$

RESULTS AND DISCUSSION

In paper [8] the theoretical treatment of transport processes and reaction in Witten-Sander clusters which are the fractal objects, is considered. Within the framework of this treatment the expression for the average probability of the survival of random walk Φ(t) after time t is received [8]:

$$\Phi(t) \sim \exp(-ct^{\delta}) ,\tag{5}$$

where c is a constant and exponent δ is equal to:

$$\delta = \frac{d_s}{2} .\tag{6}$$

Let's consider the application of this model, the physical sence and the definition of parameters, included in it, conformably to the reaction of the thermo-oxidation degradation in solid-state polymer. We consider the structure of amorphous polymer or the structure of amorphous phase of semicrystalline polymer as a set of cluster WS [7], and the oxygen molecules – as the particles making random walk on this fractal object. In such a treatment " the survival" of a particle means that it does not take part in the process of the thermo-oxidation degradation and, hence, the less $\Phi(t)$ is, the more particles (the oxygen molecules) have taken part in the process of degradation and the final MM of polymer subjected to thermal aging is lower. Or else, we assume that the only mechanism of thermal aging is the thermo-oxidation degradation, and its unique result is the decrease of polymer MM. Then the experimental values of probability of survival $\Phi(t)$ can be estimated so:

$$\Phi(t)^e = \frac{MM_i}{MM_0},\qquad(7)$$

where MM_i is molecular weight of a sample after some interval of aging, and MM_0 is molecular weight of an initial sample.

Korelman [13] showed that if the reaction proceeds according to the fractal laws (on a fractal object), its rate k will be a function of time t:

$$k \sim t^{-h},\qquad(8)$$

where h measures the degree of the local heterogeneity.

We can estimate the rate of the thermo-oxidation degradation on the reduction of polymer MM during thermal aging. In Fig. 1 the dependences of MM on the duration of thermal aging t_{ag} for PAr and HDPE+Z are given, from which their nonlinearity and, hence, the change of the degradation rate with time follows. Let's note that in case of classical behaviour (reaction on an Euclidean object) h=0 and k=const.

In Fig. 2 the dependence k on t_{ag} for PAr and HDPE+Z in double logarithmic coordinates is given. The value k was determined in this way:

$$k = \frac{MM_0 - MM_i}{t_{ag_i}}.\qquad(9)$$

From the data of Fig. 2 the linearity of plots ln k (ln t_{ag}) follows, it allows to determine an exponent h from their slope according to relationship (8). The value h is equal to 0.255 for PAr and $\sim$ 0.465 for HDPE+Z. Further it is possible to determine the effective spectral dimension d_s' from equation [13]:

$$d_s' = 2(1 - h).\qquad(10)$$

Let's note one important feature of the determination of the spectral dimension and the reason for using namely d_s', instead of d_s in the further calculation. The last parameter can be determined both theoretically [5] and experimentally [14]. For example, it can be calculated from equation [5]:

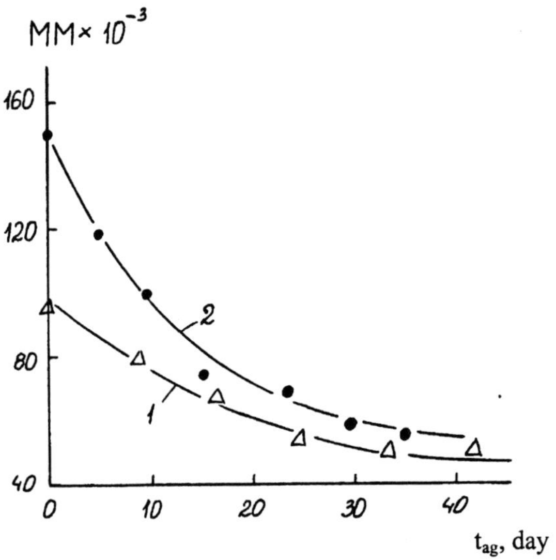

Fig. 1. Dependences of molecular weight MM on duration of aging t_{ag} for PAr (1) and HDPE+Z (2).

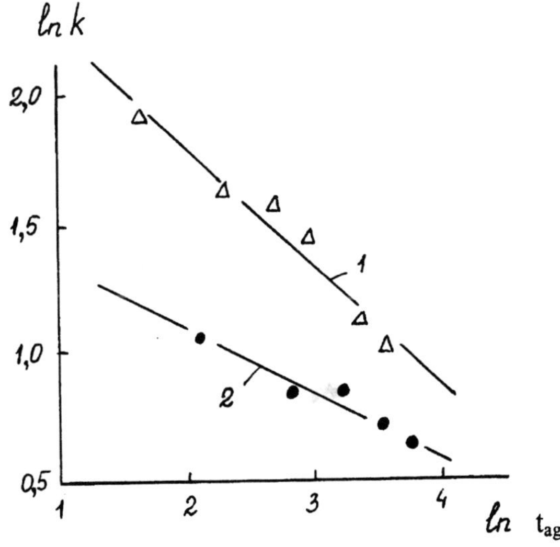

Fig. 2. Dependences of rate of the thermo-oxidation degradation k on duration of aging t_{ag} in double logarithmic coordinates for PAr (1) and HDPE+Z (2).

$$d_s = \frac{2d_f}{1+d_f},\qquad\qquad(11)$$

where d_f is a fractal dimension of polymer structure.

For cluster WS d_f=2.5 [5], for the structure of various polymer d_f=2.6-2.7 [3]. It gives an interval of values d_s for PAr and HDPE+Z ~ 1.43-1.46 according to equation (11) that is well corresponded to the results of computer modeling [5] and experimental estimations [14]. However, as thermal aging proceeds at rather high temperatures (close to temperatures of melting or glass transition temperatures of polymers), this circumstance can change value d_s owing to the presence of equation (10) it gives values d_s'$\approx$1.49 for PAr and d_s'$\approx$1.07 for HDPE+Z. Thus, if for PAr the approximate equality d_s and d_s' is observed, for HDPE+Z the difference of these parameters is rather great. It corresponds to the theorem of subordination [13]: d_s'=βd_s, where β is a parameter describing the distribution of relaxation times. For PAr $\beta\approx$1, for HDPE+Z $\beta\approx$0.74. Hence, the given example shows that the use of dimension d_s, not taking into account of the temperature changes β, can result in the essential errors.

Now, using relationship (5), in which the constant was determined by a principle of the best fitting, and the mentioned above values d_s', we can calculate the theoretical dependence $\Phi(t_{ag})^t$ and compare it with the estimation of $\Phi(t_{ag})^e$ according to equation (7). Such comparison for PAr and HDPE+Z is given in Fig. 3, from which the good correspondence between the theory and the experiment follows.

And, at last we shall consider the influence of a space type (fractal or Euclidean), in which thermo-oxidation degradation proceeds. For Euclidean space d_s'=d_s=3, as it was mentioned above. The estimation of $\Phi(t_{ag})^t$ for this case is also given for PAr in Fig. 3 (curve 5) and the comparison of curves $\Phi(t_{ag})^t$ at d_s'=1.49 (fractal space) and d_s'=3 (the Euclidean space) shows that in the latter case the decrease $\Phi(t_{ag})^t$ or the decrease of MM proceeds approximately twice as fast, than in the first, case as it was expected from the stated in the beginning of this paper reasons.

CONCLUSIONS

The results of the present paper have shown a correctness of the application of a fractal model based on the use of cluster WS for the identification of polymer amorphous state structure. As it was expected from the most common reasons, this reaction proceeds much more slowly in a fractal space, than in an Euclidean space. In essence, the represented treatment is the fractal model of structural stabilization of polymers.

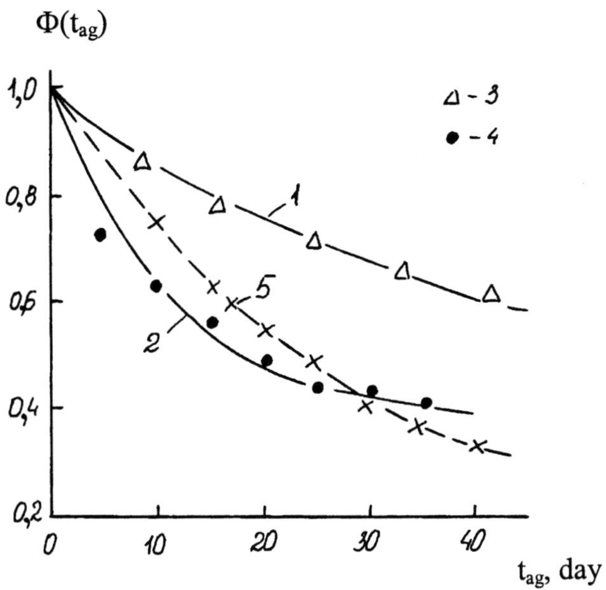

Fig. 3. Dependences of experimental $\Phi(t_{ag})^e$ (1, 2) and theoretical $\Phi(t_{ag})^t$ (3, 4) probability of survival of oxygen molecules on duration of aging t_{ag} for PAr (1, 3) and HDPE+Z (2, 4). A curve 5 is calculation $\Phi(t_{ag})^t$ for PAr at d_s'=3.0 (the Euclidean space).

REFERENCES

1. G.P.Gladyshev, Yu.A.Ershov, O.A.Shustova *Stabilization of thermo-resistant polymers.* Moscow, Khimia, 1979.

2. N.M.Emanuel, A.L.Buchachenko *Chemical Physics of aging and stabilization of polymers.* Moscow, Nauka, 1982.

3. G.V.Kozlov, V.U.Novikov *Synergetics and fractal analysis of cross-linked polymers.* Moscow, Klassika, 1998.

4. R.Rammal, G.Toulouse *J. Physiq. Lettr.,* 1983, v. 44, No 1, p. L 13.

5. P.Meakin, H.E.Stanley *Phys. Rev. Lett.,* 1983, v.51, No 16, p. 1457.

6. S.Alexander, R.Orbach *J. Physiq. Lettr.,* 1982, v. 43, No 17, p. L 625.

7. G.V.Kozlov, V.A.Beloshenko, V.N.Varyukhin *Ukrainian fizicheskii zhurnal,* 1998, v. 43, No 3, p. 322.

8. M.Sahimi, M.McKarnin, T.Nordahl, M.Tirrell *Phys. Rev. A,* 1985, v. 32, No 1, p. 590.

9. N.I.Mashukov, O.A.Vasnetsova, G.V.Kozlov, A.B.Kesheva *Lakokrasochnyi materiali i ich primenenije,* 1990, No 5, p. 38.

10. N.I.Mashukov, A.K.Mikitaev, G.P.Gladyshev, V.N.Belousov, G.V.Kozlov *Plast. massy,* 1990, No 11, p. 21.

11. V.N.Shogenov, M.A.Gazaev, M.Sh.Kardanov *In book: "Polycondensation processes and polymers"*. Nalchik, KBSU, 1987, p. 8.

12. A.A.Askadsky *Physical chemistry of polyarylates*. Moscow, Khimia, 1968.

13. P.W.Klymko, R.Kopelman *J. Phys. Chem.*, 1983, v. 87, No 23, p. 4565.

14. A.Boukenter, E.Duval, H.M.Rosenberg *J. Phys. C: Solid State Phys.*, 1988, v. 21, No 15, p. L 541.

THE PHYSICAL REASONS OF HOMOGENEOUS AND NONHOMOGENEOUS REACTIONS OF HALOIDCONTAINING EPOXY POLYMERS CURING

*G. V. Kozlov, A. A. Bejev, and G. E. Zaikov**
Kabardino-Balkarian State Agricultural Academy,
1a Tarchokov st., Nalchik, 360000, KBR, Russia

*Institute of Biochemical Physics Russian Academy of Sciences,
Kosygin st., 4, 117334, Moscow, Russia

Abstract

It has been shown that the homogeneous reaction of cross-linked polymers curing proceeds in medium with small density fluctuations and the nonhomogeneous one − in medium with large density fluctuations. Nonhomogeneous curing reaction can be simulated as proceeding in a fractal space. In this case its rate is sharply reduced in comparison with the similar reaction in an Euclidean space.

INTRODUCTION

In paper [1] it was shown that the curing mode of various haloidcontaing epoxy polymers can be of two types: homogeneous and nonhomogeneous. The first of the mentioned modes is characterized by a constant structure of formed in curing process clusters (microgels), described by it fractal dimension D, in an interval of conversion degrees of reaction $\alpha \approx 0 \div 0.7$. The monotonous increase of D is a characteristic for a nonhomogeneous mode in process of increase of reaction time in the same interval α. Such a difference is reflected in the form of the kinetic curves $\alpha(t)$. In paper [1] it was shown that the reason of such difference is the certain distribution of values D in case of a nonhomogeneous reaction and monodispersion of the mentioned values − in case of homogeneous reaction. The authors of paper [1] have made the assumption that the physical basis of presence of two mentioned types of curing reactions is the level of density fluctuations in reaction medium. If the reacting substances are distributed equipartitional (fluctuations of density are small), then in any point of reaction medium the identical conditions of formation of microgels with monodispersed value D are formed. If the reacting substances are distributed non-uniformly (fluctuations of density are great), the different ratios from curing agent and olygomer in the different points of reaction medium result in formation microgels with unsufficient or extra degree of curing, that finally is expressed in different microgels structure and polydisperse distribution D. The purpose of the present paper is the check of this assumption with the use of the scaling [2] and fractal [3] approaches on an example of two haloidcontaining epoxy polymeres which showed homogeneous (2DPP+HCE/DDM) and nonhomogeneous (EPS-1/DDM) modes of curing.

EXPERIMENTAL

Kinetics of curing of haloidcontaining olygomer on the basis of hexachlorobenzene has been investigated. This olygomer (conditional designation EPS-1) is cured by 4,4'-diaminodiphenylmethane (DDM) at stechiometric ratio DDM:EPS-1 [4].

The curing kinetics of system EPS-1/DDM was studied by a method of inverted gas chromatography [4]. The basic parameter received from processing of the experimental data, was a reaction rate constant k_r determined for interval $\alpha=0.1\div0.7$ of kinetical curve degree of conversion-time (α-t). For determination k_r the standard procedure is used: the dependence α on time of reaction t as $\lg[\alpha/(1-\alpha)]=f(t)$ which appeared to be linear were constructed. Then the value k_r was determined from a slope of these linear diagrams. Ketones (metyl ethyl ketone, 1,4-dioxsane, cyclohexanone) were chosen as the standard substances for the determination of keeping time and argon was chosen as the gas-carrier.

Besides, kinetics of curing of haloidcontaining olygomer on the basis of diphenylolpropane and hexachloroethane is studied. This olygomer (code designation 2DPP+HCE) is also cured by DDM at stechiometric ratio DDM:2DPP +HCE.

Study of curing kinetics of system 2DPP+HCE/DDM was carried out by a method IR-spectroscopy by spectrometer Perkin-Elmer. In order that the quantitative measurements would not depend on thickness of a olygomer layer put on a substrate, we applied a method of the internal standard. As a measure of the contents of epoxy groups we accepted not the optical density of analytical band 920 cm^{-1}, but its ratio to the optical density of a standard, as which the IR-band of skeletal vibrations for aromatic ring 1510 cm^{-1} is used, as this concentration was constant in process of curing. The optical densities of an analytical band and bands of the standard were determined by a method of a baseline [4].

The curing temperature T_{cur} of both studied systems was identical and equal to 393 K. It this case the calculated value D was equal to 1.78 for system 2DPP+HCE/DDM and varied in interval $1.61\div2.38$ for system EPS-1/DDM [3].

RESULTS AND DISCUSSION

Let's consider the reaction, in which particles P of a chemical substance diffuse in medium containing random located statistical nonsaturated traps T. At contact of a particle P with a trap T the particle disappears. Nonsaturation of a trap means that the reaction P+T→T can repeat itself infinite number of times. It is usually considered that is the concentration of particles and traps is large or the reaction occurs at intensive stirring, the process can be considered as the classical reaction of the first order. The samples for studies of inverted gas chromatography were prepared by the dissolution of olygomer and curing agent in acetone, putting a mixture of their solutions on a substrate and subsequent drying. Thus, the heterogeneity of a mixture, existing in a solution, was fixed during the evaporation of the solvent and was kept in a solid-state reaction of curing. In this case it is possible to consider that the law of the concentration decrease of particles with time will look like [2]:

$$c(t) \approx exp(-At), \tag{1}$$

where A is constant, t is duration of reaction.

However, if the concentration of the random located traps is small, with necessity there exist areas of space, practically free from traps. The particles getting into these areas, can reach the traps only during rather long time and, hence, the decrease of their number in the course of reaction will be slower. The formal analysis of this problem shows that the concentration of particles falls down under the law [2]:

$$c(t) \approx \exp(-Bt^{d/(d+2)}), \tag{2}$$

being dependent on the dimension of space d (B is constant). It also means the presence of large-scale fluctuations of density (heterogeneity) in reaction medium.

If the traps can move, their mobility as though averages the influence of spatial heterogeneity, so the assumption resulting to (1) will be carried out better. In this case concentration of particles falls down under the combined law [2]:

$$c(t) \sim \exp(-At)\exp(-Bt^{d/(d+2)}). \tag{3}$$

The attempt to describe the kinetical curves $\alpha(t)$ for systems 2DPP+HCE/DDM and EPS-1/DDM within the framework of equations (1)-(3) was undertaken. In this case it was supposed that:

$$c(t) = 1 - \alpha. \tag{4}$$

In Fig. 1 the dependences $\ln(1-\alpha)$ on t, corresponding to equation (1), for systems 2DPP+HCE/DDM and EPS-1/DDM are given. As it follows from the given plots, kinetics of curing of system 2DPP+HCE/DDM is well described by the linear dependence in coordinates of Fig. 1, whereas dependence $[\ln(1-\alpha)](t)$ for system EPS-1/DDM fails to linearize. It means that the homogeneous reaction of curing of system 2DPP+HCE/DDM, described by equation (1) under the mentioned above conditions, is a classical reaction of the first order proceeding in reaction medium with the small fluctuations of density.

The attempts to linearize the dependence $(1-\alpha)$ on t for system EPS-1/DDM with the use of equations (2) and (3) have not resulted in success. So, the following assumption was made. Equation (2) describes kinetics of low-molecular substances reaction at large fluctuations of density in Euclidean space with dimension d (equal to 3 in a considered case). If we assume that the formation of fractal clusters (microgels) with dimension D defines a course of curing reaction in a fractal space with dimension D, the dimension d in equation (2) should be replaced by D. The dependence $\ln(1-\alpha)$ on $t^{D/(D+2)}$, corresponding to equation (2) with the mentioned replacement, is given in Fig. 2. In such a treatment the scaling relationship (8) gives the linear correlation and this circumstance points out that the nonhomogeneous curing reaction of system EPS-1/DDM proceeds in the conditions of large fluctuations of density in a fractal space with dimension D. It is important to note that it basically distinguishes the mentioned reaction from the reactions of fractal objects in Euclidean spaces and the principle itself is similar to the formation of structures on fractal lattices [6].

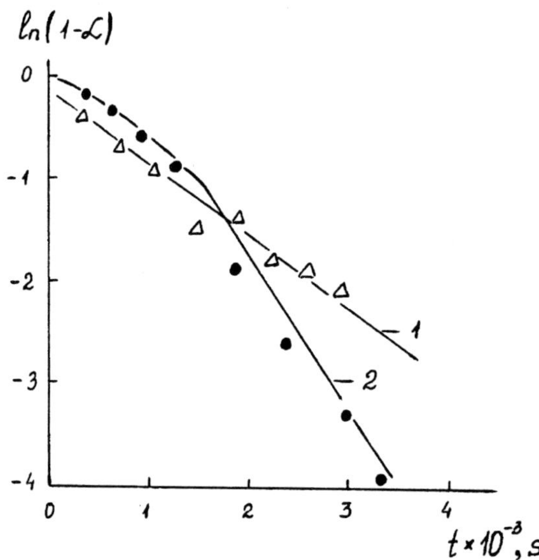

Fig. 1. Dependences $(1-\alpha)$ on reaction time t in logarithmic coordinates corresponded to the equation (1) for systems 2DPP+HCE/DDM (1) and EPS-1/DDM (2).

Using the determined values of factors A and B from the slopes of the corresponding linear plots of Fig. 1 and 2, it is possible to describe theoretically the kinetical curves $\alpha(t)$ by means of equations (1) and (2) for systems 2DPP+HCE/DDM and EPS-1/DDM, accordingly. The comparison of the calculated by the mentioned method and experimentally received curves $\alpha(t)$ for both examined systems is given in Fig. 3. As it follows from the plots shown in Fig. 3, the good correspondence to the theory and experiment is received.

Concerning the construction of the theoretical curve $\alpha(t)$ for system EPS-1/DDM it is necessary to make an important remark. As it follows from equation (2), at t=0 c(t)=1 or $(1-\alpha)=1$, i.e., at t=0 $\alpha=0$, that is obvious. However, it follows from the plot of Fig. 2 that the straight line $[\ln(1-\alpha)](t^{D/(D+2)})$ is extrapolated to $\ln(1-\alpha)=0$ (or $\alpha=0$) at finite value $t^{D/(D+2)}$, equal to ~ 12. It means that the nonhomogeneous curing reaction of system EPS-1/DDM has some period of an induction t_{in}, estimated approximately as 150 s. The physical basis of such an effect is clear. In the beginning of curing reaction (at t=0) there are not fractal clusters (microgels) in reaction mixture and it is possible to consider the available molecules of olygomer and curing agent, either as the points or as the short rods (one-dimensional objects). The formation of microgels, creating the fractal reaction space, is needs the certain time, which is the period of a theoretical curve $\alpha(t)$. So, it is necessary to use not the actual time of reaction t, but the difference $(t-t_{in})$.

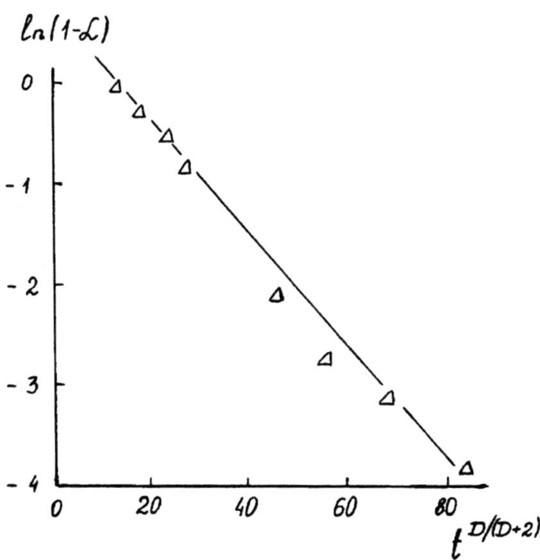

Fig. 2. Dependence $(1-\alpha)$ on parameter $t^{D/(D+2)}$ in logarithmic coordinates corresponded to the equation (2) for system EPS-1/DDM.

Let's note that the calculation of a theoretical curve $\alpha(t)$ under the condition of reaction proceeding in Euclidean space (the use in equation (2) value d=3, instead of D) gives more faster curing reaction. The curve $\alpha(t)$ for this case is shown in Fig. 3 with the broken line. It is easy to show the reasons of a various course of curves $\alpha(t)$ in Euclidean and fractal spaces in frameworks of the fractal analysis. If to consider a trajectory of a diffusive movement of olygomer and curing agent molecules as a trajectory of random walks, the number of sites $\langle S \rangle$, visited by such walks, is proportional to [7]:

$$\langle S \rangle \sim t^{d_s/2},\qquad(5)$$

where d_s is a spectral dimension of space describing its connectivity [8].

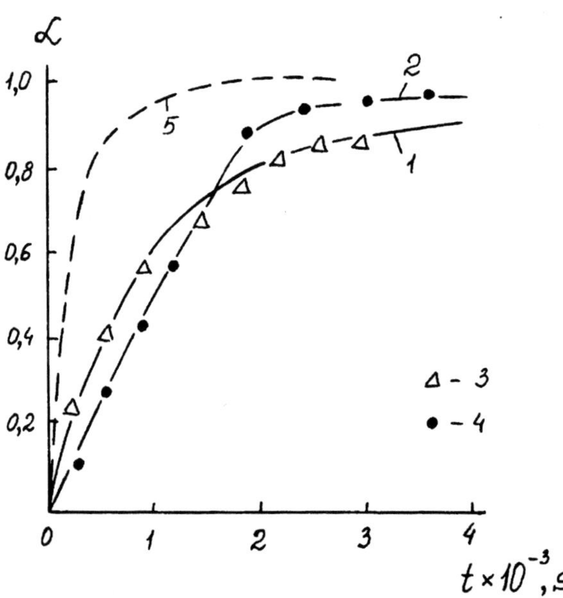

Fig. 3. Experimental (1, 2) and theoretical (3, 4, 5) kinetical curves α(t) for systems 2DPP+HCE/DDM (1, 3) and EPS-1/DDM (2, 4, 5). The curve 5 corresponds to curing reaction of system EPS-1/DDM in Euclidean space.

For an Euclidean space d_s=3 [9], for cured microgels d_s=1.33 [8]. It follows from relationship (5) that the value $\langle S \rangle$ which can be treated as a number of contacts of reacting molecules, is proportional to $t^{1,5}$ in an Euclidean space and to $t^{0,655}$ – in a fractal space. At the equal t the greater number of the pointed contacts in an Euclidean space determines faster curing reaction in comparison with a fractal space [10].

In this connection we shall note an a, interest detail. As it is shown in paper [11], for an ideal phantom network the relationship is correct:

$$\frac{D}{D+2} = \frac{d_s}{2}.$$

(6)

It is easy to see the obvious analogy between the parameters of equation (2) (at replacement d on D) and relationship (6).

CONCLUSIONS

Thus, the results of the present paper have shown that the homogeneous reaction of cross-linked polymers curing proceeds in medium with small density fluctuations and the nonhomogeneous one – in medium with large density fluctuations. The nonhomogeneous curing reaction can be simulated in terms of the scaling equation (2) as proceeding in a fractal space. In this case its rate is sharply reduced in comparison with the similar reaction in an Euclidean space.

REFERENCES

1. Kozlov G.V., Bejev A.A., Shustov G.B., Lipatov Yu.S. VII International Conf. on Chemistry and Physico-Chemistry of Olygomers. Perm, 2000, p. 207.
2. Djordjevic, Z.B. *In book: Fractal in Physics.* Edited by Pietronero L., Tosatti E. Moscow, Mir, 1988, p. 581-585.
3. Mashukov N.I., Temiraev K.B., Shustov G.B., Kozlov G.V. *In "Papers of the 6th. Int. Workshop of Polymer Reaction Engng."* Berlin, 5-7 Oct., 1998, v. 134, p. 429-438.
4. Bejev A.A. Diss. ... doct. khim. nauk, Tashkent, THTI, 1991, 323 p.
5. Stepanov B.N. Diss. ... kand. khim. nauk, Moscow, MkhTI, 1976, 154 p.
6. Vannimenus J. *Physica D,* 1989, v. 38, № 1-3, p. 351-355.
7. Sahimi M., McKartnin M., Nordahl T., Tirrell M. *Phys. Rev. A,* 1985, v. 32, № 1, p. 590-595.
8. Alexander S., Orbach R. *J. Physiq. Lettr.,* 1982, v. 43, № 17, p. L625-L631.
9. Feder E. Fractals. Moscow, Mir, 1991, 246 p.
10. Kozlov G.V., Temiraev K.B., Afaunov V.V. *Plast. massy,* 2000, № 2, p. 23-24.
11. Hess W., Vilgis T.A., Winter H.H. *Macromolecules,* 1988, v. 21, № 8, p. 2536-2542.

A Fractal Analysis of Polymers Molecular Weight Distribution: Dynamic Scaling

Kozlov G. V., Malkanduev Yu. A., and Zaikov G. E.
Kabardino-Balkarian State University,
Chernishevsky st., 173, 360004, Nalchik KBR, Russia

Institute of Biochemical Physics Russian Academy of Sciences
Kosygin st. 4, 119991, Moscow, Russia

Abstract

The correctness of the molecular weight distribution (MWD) of poly(dimethyl diallyl ammonium chloride) type has been shown in the frameworks of dynamic distribution function of the irreversible aggregation model of cluster-cluster. The build-up of a generalized distribution curve confirms the possibility to describe the polymerization processes within the frameworks of the mentioned model and allows to predict the kinetics of MWD changes as the function of initial monomer concentration c_0 and reaction time t.

Keywords: Irreversible aggregation, poly(diethyl diallyl ammonium chloride), fractal dimension.

1. Introduction

The basic process in many practically important cases is the formation of clusters due to the aggregation of particles [1]. To such phenomena it is necessary to attribute the processes of polymerization, relate to whose their basic sence being the formation of long chain macromolecules from smaller molecules. A widely used method for the description of such systems is the determination of distribution function of the sizes of clusters, which in physics and chemistry of polymers is defined as a function of molecular weight distribution (MWD).

As it is known [2] the macromolecular coils in solution are fractals whose structure or, more precisely, spatial distribution of coil's elements is characterized by its fractal dimension D. However, despite the doubtless importance, the value D gives only restricted information concerning the process of aggregation (polymerization). First, it is a static quantity and does not describe the dynamics of the aggregation process. Secondly, the quantity D characterizes the geometrical properties of only one cluster (macromolecular coil) and cannot be used for the description of cluster set in the system. The last purpose needs the examination of the dynamic distribution function of cluster sizes $n_s(t)$, which is a number of clusters consisting of s unities forming, in time t [3].

An interval of D values for macromolecular coils in solution [2] assumes that the process of polymerization proceeds according to the aggregation cluster-cluster mechanism. The dynamic distribution function of the cluster sizes for the mentioned mechanism is examined in series of papers [3-7]. The purpose of the present paper is to find out the general laws of MWD of poly(dimethyl diallyl ammonium chloride) (PDMDAACh) within the frameworks of the dynamic distribution function for the cluster-cluster aggregation mechanism.

2. EXPERIMENTAL

PDMDAACh was synthesized in water solutions at initial monomer concentration c_0: 1.0; 2.5; 4.05 and 5.0 moles/l (0.161-0.808 on mass). The initiation was carried out by thermal decomposition of ammonium persulphate with concentration of the initiator 5×10^{-3} mole/l. The conversion degree was determined by the calorimetric method. The polymerization was carried out at 333 K.

The function of MWD for PDMDAACh was determined from the changes of a sedimentation rate according to the procedure [8] at rotor rate 6-104 r.p.m. All measurements were carried out at temperature 303 K in solutions 1N NaCl. The curves MWD for PDMDAACh at various values c_0 are given in Fig. 1.

Value D was determined from equation [9]:

$$D = \frac{3}{1+a},\tag{1}$$

where a is exponent in the Mark-Houwink equation, equal for PDMDAACh 0.82 [8].

3. RESULTS AND DISCUSSION

The dynamic scaling description of the cluster size distribution of the in the diffusion-limited aggregation model cluster-cluster has been studied in some works [3-6]. This man be applied for to describe the radical polymerization PDMDAACh description for the following reason. Value $D=1.65$ ($a=0.82$ [8]) estimated on from equation (1) is a typical fractal dimension for aggregates forming in the diffusion-limited cluster-cluster aggregation process [1]. The applicability of this model for the description of polymerization repeatedly proved to be true [7, 10]. On this base for the description of MWD for PDMDAACh dynamic scaling function man be used as [3]:

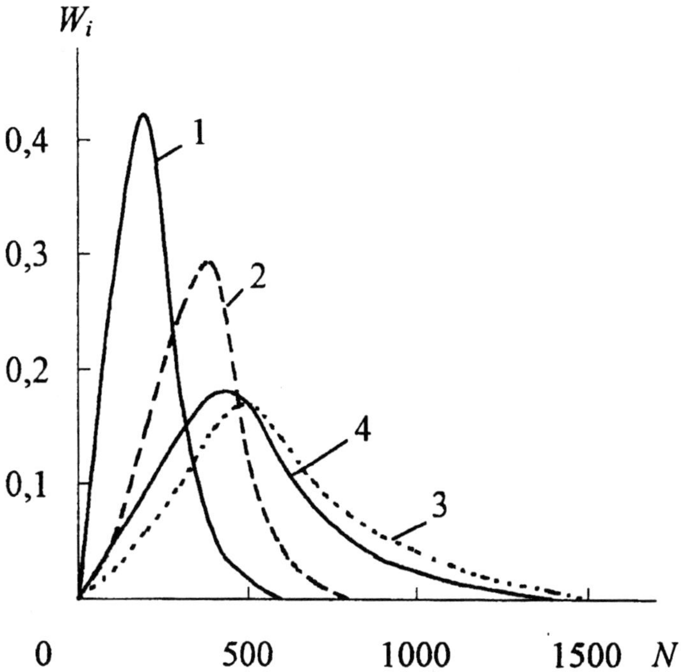

Fig. 1. Curves MWD for PDMDAACh, synthesized at c_0: 1.0 (1), 2.5 (2), 4.05 (3) and 5.0 (4) moles/l.

$$n_s(t) \sim S^{-2}t^z, \tag{2}$$

where t is reaction time accepted the equal gelation time, z is an exponent.

Consider the methods of the estimation of parameters entering equation (2). The values $n_s(t_g)$ where t_g is gelation time, are determined from the experimental curves MWD $W_i(N)$ (Fig. 1) as follows. We assume that $N=S$ and that a fraction of polymer corresponding to S_i is equal to a summary weight fraction of polymer W_i. Dividing W_i in S_i and multipled man be playing received by the maximum degree of conversion Q, receive value $n_s(t_g)$ or n_i for S_i. The exponent z is calculated according to equation [11]:

$$z = D(1-\alpha)-(d-2), \tag{3}$$

where α is an exponent in the relationship for diffusion constant D_s as function S ($D_s \sim S^\alpha$) and d is an Euclidean space dimension, in which the fractal, equal, in our case, to 3, is considered. The value D_s can be calculated as follows [12]:

$$D_s = \frac{kT}{6\pi\eta_0 R_g \beta}, \tag{4}$$

where k is a Boltzmann constant, T is testing temperature, η_0 is initial viscosity of reactionary mixture, R_g is a radius of gyration of diffusing particle (macromolecular coil), β is a numerical coefficient defined by boundary conditions of particle surface.

In its turn, as it was mentioned above, value D_s within the frameworks of the irreversible aggregation models usually is accepted as [3, 11]:

$$D_s \sim S^\alpha. \tag{5}$$

From the comparison of relationships (4) and (5) it is easy to receive:

$$\alpha \sim -\ln \eta_0. \tag{6}$$

The values η_0 for the monomer solutions, used at synthesis PDMDAACh, are given in paper [13].

As it was pointed out [3, 11], the value α defines a form of a distribution function. For $\alpha < -0.5$ the "bell"-shaped distribution curves having a maximum are received. For $\alpha > -0.5$ the monotonously decaying curves are typical. As follows from Fig. 1, the experimental curves MWD have the shape with maximum. On the basis of absolute values η_0, we have assumed in relationship (6) a sign of the equality and in this case values α vary whithin limits $\sim$ 1-3 (values η_0 are given in relative units [13]) and this interval is completely meets the shape of experimental curves MWD (Fig. 1). Besides, as it was shown earlier [14], the position of maximum of the curve MWD is determined by the relationship:

$$N_{\max} \sim \frac{-\alpha}{1-\alpha}. \tag{7}$$

It is not difficult to see that the increase of c_0, giving the increase η_0, is defines the build-up of absolute value α and increase $N_{\max}$, corresponded with of Fig. 1.

In Fig. 2 the dependencies $(S^2 n_s)$ on (St^z) in the double logarithmic coordinates for four polymers PDMDAACh, synthesized at various c_0, are shown. As follows from the data of this Figure, all four curves MWD, shown in Fig. 1, are described by the unique curve. This result confirms the correctness of use the application of the irreversible aggregation models for the description of the polymerization process.

Using generalized curve shown in Fig. 2, it is possible to predict theoretically the kinetics of MWD changes as a function of time. For this purpose, at first, we are set by values $S=N$ and determine value St^z at arbitrary t. Then, from the diagram of Fig. 2 the value $S^2 n_s$, is found corresponding to it, and according to the described above procedure defined W_i. Then the same calculation is iterated for another S, etc. In Fig. 3 as an example, the comparison of curves MWD for c_0=4.0 moles/l is given at $t=t_g$=108 min. and t=53 min. As is seen from the comparison of curves 1 and 2 in Fig. 3, the decrease t ($t<t_g$) gives in bias of a maximum MWD to lower the molecular weight values and narrowed MWD. Besides, in Fig. 3 the curve MWD for c_0=1.0 moles/l is given at $t=t_g$ (curve 3) and from the comparison of all three curves it follows that the decrease t and c_0 give the similar effect.

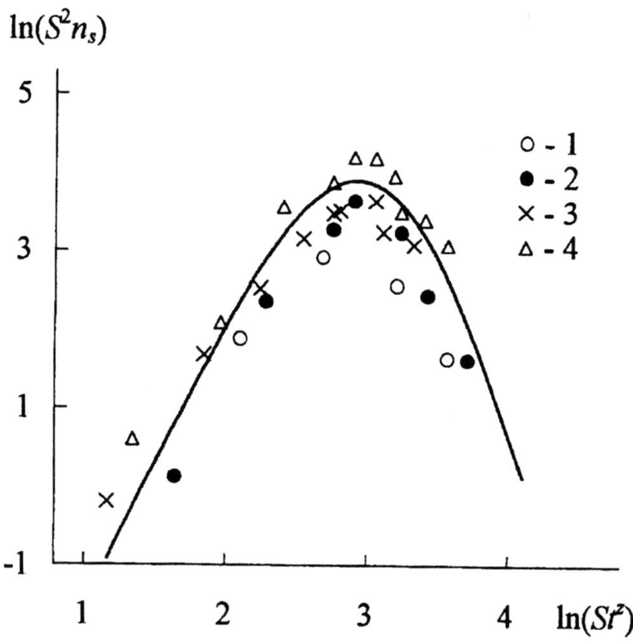

Fig. 2. Generalized MWD curve in the form of functions $S^2 n_s(St^2)$ in double logarithmic coordinates for PDMDAACh, synthesized at c_0: 1.0 (1), 2.5 (2), 4.05 (3) and 5.0 (4) moles/l.

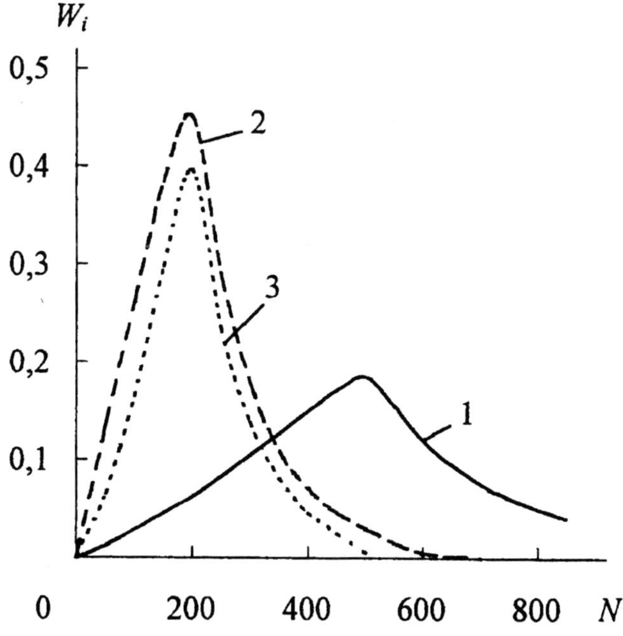

Fig. 3. Experimental curves MWD at gelation time for PDMDAACh, synthesized at c_0=4.0 (1) and 1.0 (3) moles/l. Calculated curve MWD at t=53 min. for PDMDAACh, synthesized at c_0=4.0 moles/l (2).

4. CONCLUSIONS

Thus, the results of the present paper have shown the correctness of the description the polymer of molecular weight distribution on an example of poly(dimethyl diallyl ammonium chloride) within the frameworks of dynamic distribution function of the irreversible aggregation cluster-cluster model. It is proved by the build-up of a generalized distribution curve and allows to predict the kinetics of MWD change as the function of monomer initial concentration c_0 and reaction time t.

REFERENCES

1. Kokorevich A.G., Gravitis Yu.A., Osol-Kalnin V.G. *Khimia drevesiny* 1989; 1: 3-24.
2. Baranov V.G., Frenkel S.Ya., Brestkin Yu.V. *Dokl. AN SSSR* 1986; 290(2): 369-372.
3. Meakin P., Vicsek T., Family F. *Phys. Rev. B* 1985; 31(1): 564-569.
4. Vicsek T., Family F. *Phys. Rev. Lett.* 1984; 52(19): 1669-1672.
5. Kolb M., Herrmann H.J. *J. Phys. A: Math. Gen.* 1985; 18(8): L435-L441.
6. Kolb M. *Physica A* 1986; 140(1-2): 416-420.
7. Shiyan A.A. *Vysokomolek. Soed. B* 1997; 37(9): 1578-1580.
8. Timofejeva G.J., Pavlova S.A., Wandrey Ch., Jaeger W., Hahn M., Linow K.-Y., Gornitz E. *Acta Polymerica* 1990; 41(9): 479-484.
9. Karmanov A.P., Monakov Yu.B. *Vysokomolek. Soed. B* 1995; 37(2): 328-331.
10. Novikov V.U., Kozlov G.V. *Russ. Adv. In Chemistry* 2000; 69(6): 572-599.
11. Kolb M. *Phys. Rev. Lett.* 1984; 53(17): 1653-1656.
12. Shamurina M.V., Roldugin V.I., Prjamova T.A., Vysotzky V.V. *Koll. Zh.* 1994; 57(4): 580-584.
13. Topchiev D.A., Malkanduev Yu.A., Korshak Yu.V., Mikitaev A.K., Kabanov V.A. *Acta Polymerica* 1985; 36(7): 372-374.
14. Botet R., Jullien R. *J. Phys. A* 1984; 17(12): 2517-2530.

A FRACTAL FORM OF THE MARK-KUHN-HOUWINK EQUATION

Kozlov G. V., Dolbin I. V., and Zaikov G. E.
Kabardino-Balkarian State University,
Chernishevsky st., 173, 360004, Nalchik KBR, Russia

Institute of Biochemical Physics Russian Academy of Sciences
Kosygin st. 4, 119991, Moscow, Russia

Abstract
The possibility of deriving the Mark-Kuhn-Houwink equation in terms of fractal analysis concepts was shown. The constant K_η in this equation was found to depend on the fractal dimensionality D and the molecular mass of the monomer unit characterizing the macromolecular coil structure and the chemical structure of a polymer, respectively.

The Mark-Kuhn-Houwink equation derived from the analysis of a great body of experimental data is widely used for determining M_η from the measured intrinsic viscosity $[\eta]$ of polymer solutions [1], This equation has the form

$$\left[\eta\right] = K_\eta M_\eta^{a_\eta},\tag{1}$$

where $K\eta$ and $a\eta$ are constants characterizing a polymer in a given solvent at a given temperature in a certain molecular mass range.

Analysis of experimental data revealed [2] that the K_η and a_η constants involved in equation (1) correlate with one another for many flexible-chain polymers dissolved in good solvents. Thus, it was found that

$$\left[\eta\right] \sim \frac{21}{m_0}\left(\frac{4\times10^{-4}M}{m_0}\right)^{a_\eta}\tag{2}$$

where m_0 is the molecular mass of a monomer unit.

In some cases, the use of empirical relationships such as equation (2) enables molecular mass evaluation to be performed without detailed study of the samples. A more exact relationship for K_η was proposed by Budtov [1]:

$$K_\eta = \frac{0,46}{m_0} \frac{A^2 l \times 10^{24}}{\left(7,3 S m_0\right)^{a_\eta}}$$

(3)

Here, A is the Kuhn segment length, l is the length of a monomer unit, and S is the number of monomer units per segment.

For polymers with the same chain rigidity, equation (3) is very close to empirical formula (2). Relationships (1)-(3) were noted [1] to be very useful in the analysis of solitary polymer samples when they have not been characterized in a relevant solvent but their structural parameters are known.

The application of fractal analysis demonstrates [3] that the constant a_η is not an empirical quantity but characterizes the-structure of a macromolecular coil in solution and is related to the fractal dimension D of the coil by the simple relationship

$$a_\eta = \frac{3-D}{D}$$

(4)

This circumstance allows a fractal variant of the Mark-Kuhn-Houwink equation to be derived with the use of the well-known relationships between $[\eta]$, the molecular mass, and the radius of gyration R_g for a macromolecular coil. This is what we have done in the present work.

We used the a_η and K_η values reported in [4] for solutions of various polyarylates in three solvents (1,1,2,2-tetrachloroethane, tetrahydrofurane, and 1,4-dioxane). In addition, we used the $[\eta]$ dependence on the solubility parameter δ_s of solvent for the polyarylate based on isophthalic acid and phenolphthalein (F-1) [5].

For the derivation of the fractal variant of the Mark-Kuhn-Houwink equation, we started from the well-known formula [1]

$$[\eta] = 6^{1/2} \Phi(\alpha) \frac{\langle R_g^3 \rangle}{M},$$

(5)

where Φ is the form factor of a macromolecular coil under rotational friction and a is the swelling ratio of the coil.

For the sake of convenience, a is adopted to be equal to the viscous swelling ratio defined as [1]

$$\alpha_\eta^3 = [\eta]/[\eta]_\theta,$$

(6)

where $[\eta]$ and $[\eta]\theta$ are the intrinsic viscosities of a polymer in arbitrary and θ solvent, respectively.

The simplest relationship between Φ and α is [1]

$$\Phi(\alpha) = \Phi_\theta\left(0,753 + \frac{0,247}{\alpha^3}\right), \tag{7}$$

where $\Phi\theta$ is Φ for a equal to unity.

The relationship between R_g and molecular mass M can be expressed as an approximate expression [6],

$$R_g = BN^{1/D} = \left(\frac{M}{m_0}\right)^{1/D} \tag{8}$$

where B is the proportionality coefficient and N is the degree of polymerization. The value of B can be determined with the following, relatively simple procedure. For F-2 polyarylate dissolved in 1,1,2,2-tetrachloro-ethane, the Mark-Kuhn-Houwink equation has the form [4]

$$[\eta] = 2,421 \times 10^{-4} M^{0,696} \tag{9}$$

Taking an arbitrary value of 8×10^4 for M, we obtain from equation (9) $[\eta]=0.626$ dl/g for F-2. Then, for these M and $[\eta]$ values and $\Phi(\alpha) = 2 \times 10^{23}$ [1], equation (5) yields $R_g \approx 234$ Å. Finally, the B value determined from equation (8) at $m_0=440$ appears to be 12.4 Å. Values of B for other polyarylates are determined in a similar way.

Combining equations (5)-(8), we can obtain a relationship between $[\eta]$ and M, which is similar to the Mark-Kuhn-Houwink equation:

$$[\eta] = c(\alpha)\frac{M^{(3-D)/D}}{m_0^{3/D}}, \tag{10}$$

where c is some constant depending on α, because it contains the parameter $\Phi(\alpha)$. This makes it possible to evaluate the coefficient $c(\alpha)$ at varying a in accordance with equation (7). The value $\alpha=\alpha_\eta$ itself can be evaluated by equations (1), (2), and (6), if $[\eta]_\theta$ is determined at $a_\eta=0.5$ (or $D=2.0$) [7].

It is known [8] that the ratio $([\eta]/[\eta]_\theta)$ is a single-valued function of D, as follows from the equation

$$D = \frac{5\left([\eta]/[\eta]_\theta\right)^{2/3} - 3}{3\left([\eta]/[\eta]_\theta\right)^{2/3} - 2} \tag{11}$$

Thus, expression (10) suggests that the factor K_η depends only on two parameters: D and m_0. A similar empirical dependence of $[\eta]$ on M is reported in [9]. The final equation for calculating the K_η value takes on the form

$$K_\eta = \frac{8,1(0,753 + 0,247/\alpha^3)}{m_0^{3/D}} \qquad (12)$$

The coefficient K_η as defined in equation (12) is expressed in dl/g and g/mol for $[\eta]$ and m_0, respectively. The factor 8.1 in this equation is chosen for the best fit to the experiment, because variations in B are possible for various polyarylates.

Figure 1 compares theoretical K_η values K_η^{calc} calculated by equation (12) with corresponding experimental values K_η^{exp} [4] for polyarylates. This plot demonstrates good agreement between K_η^{calc} and K_η^{exp}. In this case, D values were calculated by equation (4).

It is known [10] that the D value is determined by interactions between elements of macromolecular coils and by polymer – solvent interactions. The lack of a theory of regular solutions complicates accurate description of the second type of interaction.. So, we will use the following approximate expression to determine D) [11]:

$$D \approx 1,5 + 0,35(\Delta\delta_f)^{2/(1+\delta_c)}, \qquad (13)$$

which was obtained in the framework of two-component model for the solubility parameter [12]. The component δ_f characterizes the energy of dispersion interactions and interactions of dipolar bonds, and δ_c characterizes the energy of interaction between electron-deficient and electron-rich atoms. The values of δ_f and δ_c as borrowed from [12] and together with D values calculated by equation (13) are listed in the table. According to Wiehe [12], the solubility parameter δ is expressed as

$$\delta^2 = \delta_f^2 + \delta_c^2, \qquad (14)$$

while the $\Delta\delta f$ value is defined as

$$\Delta\delta_f = \left| \delta_f^p - \delta_f \right|, \qquad (15)$$

where δ_f^p is the δf value for a polymer (for F-1 polyarylate δ_f^p =18 (MJ/m3)1/2 [12]).

Figure 2 compares function $[\eta](\delta_s)$ as given by the experimental data [5] (solid line) with the $[\eta]$ values calculated according to equation (10) (data points) for F-1 polyarylate. The polymer molecular mass was calculated by equation (1) using the published values of $[\eta]$ [5],

α_η, and K_η [4] for an F-1 solution in 1,1,2,2-tetrachlo-roethane. It is obvious that the data calculated by equation (10) fit well to the experimentally obtained curve.

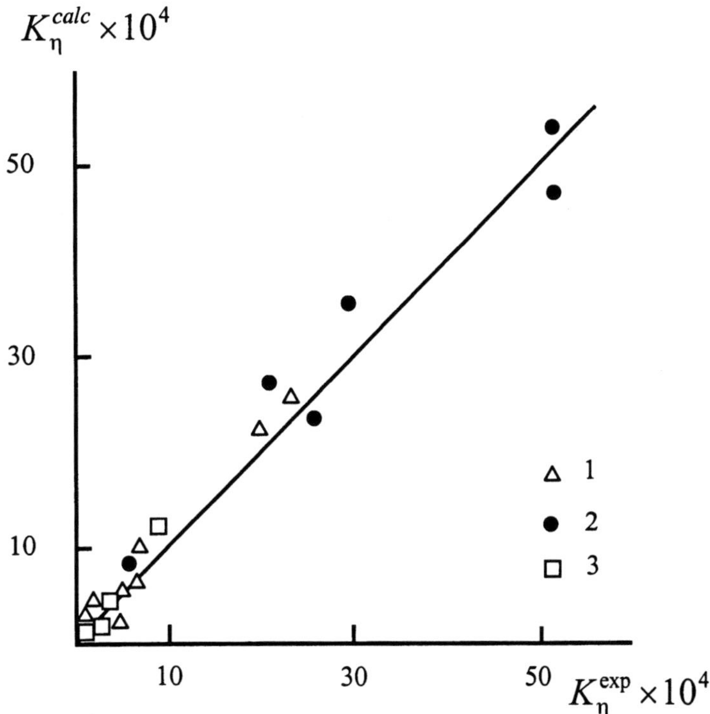

Fig. 1. Correlation between experimental $\left(K_\eta^{exp}\right)$ Mark-Houwink constants and those calculated by equation (12) K_η^{calc} forthe solutions of polyarylates [4] in (1) 1,1,2,2-tetrachloroethane, (2) tetrahydrofurane, and (3) 1,4-dioxane.

In summary, the results presented in this work demonstrate a possibility of theoretical derivation of the form of the Mark-Kuhn-Houwink equation, which is based on the concepts of fractal analysis, and also the prediction of intrinsic viscosity [η] for dilute solutions. The constant K_η in the Mark-Kuhn-Houwink equation depends on the fractal dimension D characterizing the structure of a macromolecular coil and on the molecular mass m_0 of a monomer unit.

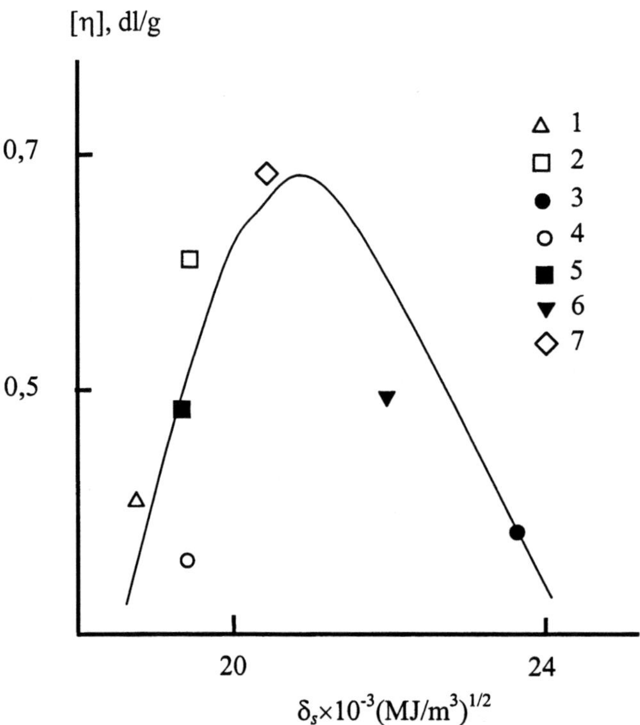

Fig. 2. Experimental (solid curve) and theoretical (data points) intrinsic viscosity [η] vs. the solubility parameter δ_s of a solvent for polyarylate F-1 dissolved in (1) cyclohexanone, (2) chloroform, (3) N,N-DMFA, (4) THF, (5) methylene chloride, (6) N,N-DMA, and (7) dichloroethane.

Table

Published and calculated characteristics of F-1 polyarylate solutions in different solvents

Solvent	δ_f, $(MJ/m^3)^{1/2}$, [12]	δ_c, $(MJ/m^3)^{1/2}$, [12]	D
Cyclohexanone	17,36	12,31	1,770
Tetrahydro furan	16,79	9,82	1,806
N,N-Dimethylformamide	17,40	16,64	1,783
Chloroform	17,66	6,15	1,675
Methylene chloride	18,44	8,17	1,733
N,N-Dimethylacetamide	17,36	13,60	1,776
Dichloroethane	18,17	9,68	1,656

REFERENCES

1. Budtov, V.P., *Fizicheskaya khimiya rastvorov polimerov (Physical Chemistry of Polymer Solutions),* St. Petersburg: Khimiya, 1992.
2. Frenkel', S.Ya., *Vvedenie vstatisticheskuyu teoriyupoli-merizatsii (Introduction to the Statistical Theory of Polymerization),* Moscow: Nauka, 1965.
3. Karmanov, A.P. and Monakov, Yu.B., *Polymer Science,* Ser. B, 1995, vol. 37, no. 2, p. 63.
4. Askadskii, A.A., *Fizikokhimiya poliarilatov (Physical Chemistry of Polyarylates),* Moscow: Khimiya, 1968.
5. Askadskii, A.A., *Struktura i svoistva teplostoikikh polimerov (Structure and Properties of Heat-Re si slant Polymers),* Moscow: Khimiya, 1981.
6. Kozlov, G.V., Temiraev, K.B., and Afaunov, V.V., *Available from VINITI,* 1998, Moscow, no. 10.
7. Baranov, V.G., Frenkel', S.Ya., and Brestkin, Yu.V., *Dokl. Akad. Nauk SSSR,* 1986, vol. 290, no. 2, p. 369.
8. Kozlov, G.V., Temiraev, K.B., Ovcharenko, E.N., and Lipatov, Yu.S., *Dokl. Ukr. Akad. Nauk,* 1999, no. 12, p. 136.
9. Temiraev, K.B., Available from *VINITI,* 1998, Moscow, no. 2291.
10. Family, F., *J. Stat. Phys.,* 1984, vol. 36, no. 5/6, p. 881.
11. Shogenov, V.N., Beloshenko, V.A., Kozlov, G.V., and Varyukhin, V.N., *Fiz. Tekh. Vys. Davlenii,* 1999, vol. 9, no. 3, p. 30.
12. Wiehe, I.A., *Ind. Eng. Chem. Res.,* 1995, vol. 34, no. 2, p. 661.

TIME DEPENDENT TRANSITION FROM FRACTAL TO EUCLIDEAN KINETICS AT POLYARYLATEARYLENE SULPHONOXIDE THERMOOXIDATIVE DEGRADATION

Kozlov G. V., Batyrova H. M., and Zaikov G. E.
Kabardino-Balkarian State University,
Chernishevsky st., 173, 360004, Nalchik KBR, Russia

Institute of Biochemical Physics Russian Academy of Sciences
Kosygin st. 4, 119991, Moscow, Russia

Abstract

In the conditions of autoacceleration regime of the thermooxidative degradation process the transition from the reaction of radicals in a fractal space to the same reaction in an Euclidean space in observed. The change of a space type defines the strong change of the oxygen consumation rate, i.e., the thermooxidative degradation rate.

Keywords: Fractal space, Euclidean space, thermooxidative degradation, spectral dimension, autoacceleration regime, mobile radicals.

INTRODUCTION

As is it known [1], one of the most wide spread types of kinetic curves "amount of consumed oxygen – time" at the thermooxidative degradation of polymers $N_{O_2}(t)$ is the curves of an autoaccelerated type with the clearly expressed induction period. As a rule, the curves $N_{O_2}(t)$ of this type can be divided into three parts: the induction period, where very slow oxygen consumption happens; the period of autoacceleration, where the rate of oxygen consumation fast increases and the linear part with very high rate of oxygen consumption [1]. Within the framework of the fractal kinetics of chemical reactions the curvilinearity of dependences $N_{O_2}(t)$ supposes the passing of the reaction in a fractal space, and the linearity – in an Euclidean space [2]. Within the framework of the fractal kinetics the reactions rate k is described by the relationship (for the case of the autoacceleration) [3]:

$$k \sim t^{1-h},\tag{1}$$

where t is the duration of reaction, h is a heterogeneity exponent ($0 \leq h \leq 1$).

For Euclidean spaces the value $h=0$ and the reaction rate are constant (does not depend on the time). Applarently, the last case corresponds to the linear dependence $N_{O_2}(t)$. Thus, there is a reason for assuming that for the curves $N_{O_2}(t)$ of the autoaccelerated type the transition of space, in which the thermooxidative degradation passes, from fractal to Euclidean, is observed. Such transition is assumed in paper [4] for reactions of the type:

$$A + A \rightarrow products\ of\ reaction.\tag{2}$$

As it is known [1], the oxidation of polymers in case of the curve $N_{O_2}(t)$ of the autoaccelerated type realizes by the radical-chain mechanism with the degenerated branching, and the free radicals [5]. Thus, the type of reaction in case of the curves $N_{O_2}(t)$ of the indicated type corresponds to reaction (2). Therefore, we can use the model [4] for the description of the thermooxidative degradation characterized by the kinetic curves of an autoacselerated type. The purpose of the present papers is to determine the availability of time-dependent transition from the fractal to the Euclidean behaviour in the process of the thermooxidative degradation of block-copolymer polyarylatearylene sulphonoxide (PAASO).

EXPERIMENTAL

PAASO, obtained by the different methods of polycondensation is studied. These methods and legends of block copolymers are listed in table 1. PAASO is synthesized from diane, mixture (1:1) of dichloroanhydrides of tere- and isophalic acids and 4,4'-dichlorodiphenylsulfone with molecular weight 4600, having the following structure:

The average molecular weight $\overline{M}_w$ is determined by method of approaching to equilibrium (Archibald method) in ultracentrifuge 3170 of the corporation MOM (Hungary). The value $\overline{M}_w$ for the studied PAASO is equal to 58×10^3 [6]. The glass transition temperature T_g of studied copolymers is defined by a dielectric method. The studies are carried out by quantometer BM-560 "Tesla" at frequency 1 MHz within the temperature range 293-573 K [6]. The value T_g for the studied PAASO is equal to 483 K.

For the studies of the thermooxidative degradation processes in air the ampullary technique is used. The working volume of ampoules is equal to 3×10^{-5} l. The average initial contents of oxygen make up the value 2.5-3.0 mol. O_2/mol of polymer. The kinetic curves of the oxygen consumption $N_{O_2}(t)$ is obtained at temperatures 573 and 623 K.

RESULTS AND DISCUSSION

As it is shown in paper [5], the product which is responsible for the autoaccelerated mode of curves $N_{O_2}(t)$, can be the aldehidic groups formed from the metyl groups (which are present in PAASO, see formula of polymer) under the scheme:

$$- CH_3 \rightarrow - \bullet CH_2 \rightarrow - CH_2OO\bullet \quad \begin{cases} \rightarrow - CH_2 - O - O - H \\ \\ \rightarrow - C\overset{O}{\underset{H}{\diagup}} + \bullet OH \end{cases} \qquad (3)$$

The oxidation of the aldehydic groups realizing according the scheme:

$$R - \underset{\underset{\displaystyle \text{H}}{\diagdown}}{\overset{\overset{\displaystyle \text{O}}{\diagup\!\diagup}}{\text{C}}} + O_2 \rightarrow R - CO\bullet + HO_2\bullet \tag{4}$$

$$\downarrow$$

$$R\bullet + CO$$

results in the formation of active free radicals $R\bullet$ and low molecular radicals $HO_2\bullet$. The latter are the mobile radicals, and it allows to consider them as the particles making random walks and to use the equation (2). The standard classical reaction of order 1/2 is described by relation [4]:

$$-\frac{d\rho}{dt} \sim \rho^{1/2}, \tag{5}$$

where ρ is the density of random walks (radicals)

As it was mentioned above, for the microscopically heterogeneous (fractal) medium the relationship (5) is written as follows [4]:

$$-\frac{d\rho}{dt} \sim t^{1-h}\rho^{1/2}. \tag{6}$$

And the integrated equation for reaction rate can be written so:

$$\rho_0^{1/2} - \rho^{1/2} \sim t^{2-h}$$

where ρ_0 is the initial density of random walks.

Besides, the effective spectral dimension d_s' of the medium, in which the reaction is passing, is connected with the heterogeneity exponent h as follows [2-4]:

$$d_s' = 2(1-h).$$

For time $t=0$, the value ρ_0 in the conditions of our experiment is defined as the limiting value of oxygen $N_{O_2}^{\infty}$ which is capable to be consumed at the oxidation. As the estimation [7] have shown, for the complete oxidation of methyl and aliphatic groups of PAASO it is approximately required 24.1 mol. O_2/basis mol. of polymer. The value ρ is defined as $\left(N_{O_2}^{\infty} - N_{O_2}\right)$. In Fig. 1 the dependence $N_{O_2}^{1/2}$ on dN_{O_2}/dt for PAASO at $T=623$ K is shown. This dependence confirms the correctness of using the relationship (5).

In Fig. 2 the dependences of the defined in this way parameter $\left(\rho_0^{1/2} - \rho^{1/2}\right)$ on time t in double log-log coordinates for PAASO at $T=573$ and 623 K are shown. As it is possible to see, at small t, correspondeing to the induction period, the dependences are linear and have a small slope, from which it is possible to determine the values $(2-h)=1.04$, $h\approx0.96$ and $d_s'=0.08$. The values of parameters h and d_s' correspond to passing the thermooxidative degradation in a fractal space [2-4]. The value d_s' is connected to the fractal dimension d_f of medium, where the reaction is passing, by the equation:

$$d'_s = \frac{2(2d_f - d)}{d+2},\qquad(9)$$

where d is the dimension of Euclidean space (in our case $d=3$).

For $d_s'=0.08$ $d_f=1.60$. The last value is corresponded to the fractal dimension of macromolecular coil surface, which is defined as follows:

$$d_f = \Delta_f - 1,\qquad(10)$$

where Δ_f is fractal dimension of a macromolecular coil. The value Δ_f for PAASO is approximately equal to 2.6 [8].

Then the gradual increase of slope of the dependences shown in Fig. 2 is observed, that corresponds to the increase d_s' and the decrease h. And at last, the third part is linear again, but it has much more bigger slope appropriate to $(2-h)=2$, $h=0$ and $d_s'=2.0$. The latter means passing of the reaction in an Euclidean space [3]. Thus, the thermooxidative degradation rate is largely defined by the type of a space, in which it is passing. In a fractal space with the low level of connectivity degree the reaction rate is insignificant (induction period). The increase d_s' from ~0.08 up to ~2.0 determites the sharp increase of oxygen consumption rate (autoacceleration). And at last, the transition to an Euclidean space $(d_s'\approx2.0$ [3]) yields a linear part of the kinetic curves $N_{O_2}(t)$ with a constant and very high oxygen consumption rate.

The data of Fig. 2 are corresponded to the results of computer simulation. Argyrakis and Kopelman have shown [9], that the effective spectral dimension d_s' increases with the time from ~1.3 up to 1.9 for the two-dimensional Euclidean space and approaches to 2 much more faster for the three-dimensional space. In the considered case this effect is conditioned the availability of the clusters (macromolecular coils) of different sizes in a melt of PAASO. The availability of the clusters of small sizes determines the decrease d_s', that is taken into account by the introduction of the corrections of different kinds [9]. However, such corrections are true only for small times, and for large enough t the value d_s' begins to increase, what the graphics of Fig. 2 point out.

The termination of the induction period and the transition to the mode of autoacceleration for PAASO at the different temperatures is observed at different values of time t (Fig. 2). This difference is conditioned by the variation of temperature T and is described within the framework of the known Arrenius equation [10]:

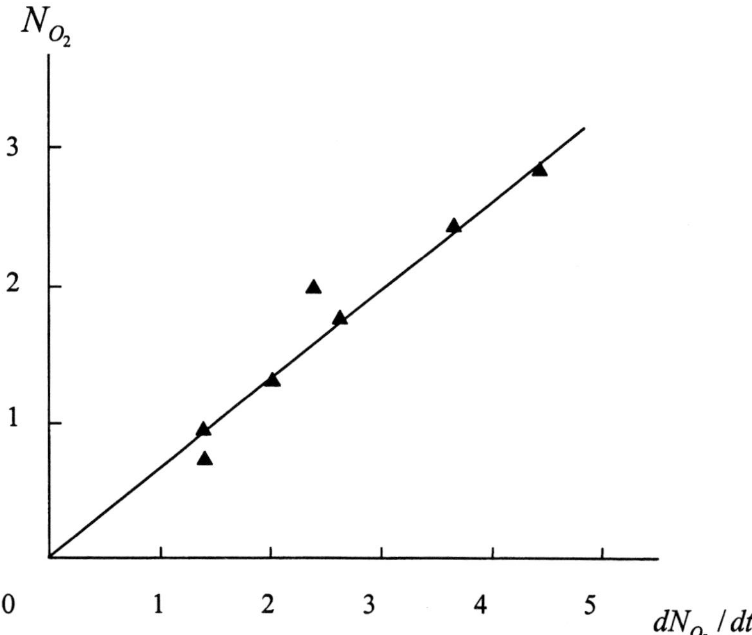

Fig. 1. Dependence of the consumed oxygen amount N_{O_2} on the reaction rate dN_{O_2}/dt for PAASO at T=623 K. This correlation is used for the determination of reaction order.

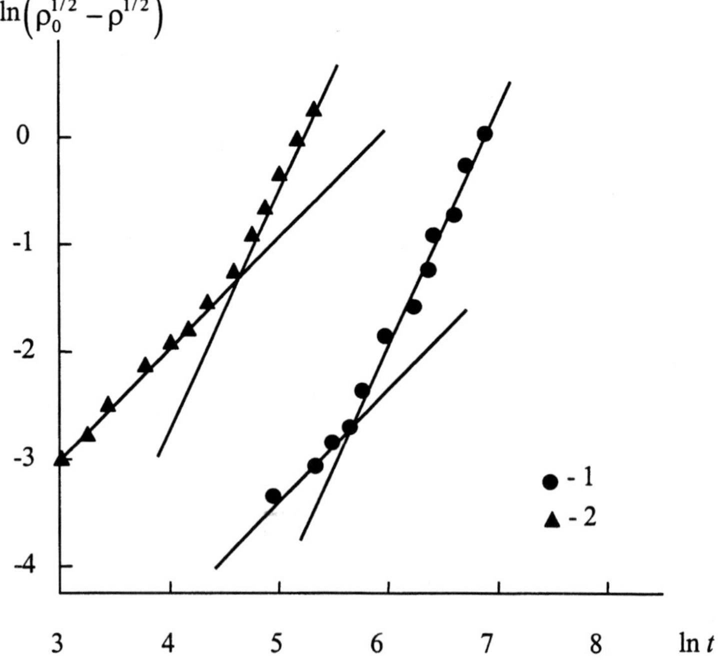

Fig. 2. Dependences of the parameter $\left(\rho_0^{1/2} - \rho^{1/2}\right)$ on time t in double log-log coordinates for PAASO at T=573 (1) and 623 K (2).

$$E_{act} = \frac{R\ln\left(t_2/t_1\right)}{\left(1/T_1\right)-\left(1/T_2\right)}, \tag{11}$$

where E_{act} is the activation energy of process, R is a universal gas constant, t_1 and t_2 are times of transitions appropriate to testing temperatures T_1=573 K and T_2=623 K.

The estimation on the equation (11) has yielded the following value of the activation energy: $E_{act}\approx$98.4 kJ/mole, which corresponds to the value, obtained for the same value E_{act} is obtained for the transition to the second linear part of dependences of Fig. 2 according to the equation (11). The correctness of Arrenius treatment (the equation (11)) means that at the temperature and time variations the mechanism of the thermiooxidative degradation of PAASO does not vary and in both cases the relationships (3) and (4) are correct.

CONCLUSIONS

The results of the present paper have shown that in the conditions of autoacceleration regime of the thermooxidative degradation process the transition from the reaction of radicals in a fractal space to the same reaction in an Euclidean space. The change of a space type defines the sharp variation of the oxygen consumation rate, i.e., the thermooxidative degradation rate, which is much more higher in an Euclidean space in comparison with a fractal space at the invariable chemical mechanism of reaction.

REFERENCES

1.   Emanuel N.M. *Vysokomolek. Soed. A,* 1985, v. 27, № 7, p. 1347-1363.
2.   Klymko P.W., Kopelman R. *J. Phys. Chem.,* 1983, v. 87, № 23, p. 4565-4567.
3.   Kopelman R. In book: *Fractals in Physics.* North-Holland, Amsterdam, Oxford, New York, Tokyo, 1986, p. 524-527.
4.   Kopelman R., Klymko P.W., Newhouse J.S., Anacker L.W. *Phys. Rev. B,* 1984, v. 29, № 6, p. 3747-3748.
5.   Shlyapnikov Yu.A., Kirjushkin S.G., Mar'in A.P. *Antioxidative stabilization of polymers.* Moscow, Khimiya, 1986, 256 p.
6.   Batyrova H.M. *Diss.... kand. khim. nauk.* Moscow, MkhTI, 1985, 160 p.
7.   Shabaev A.S. *Diss.... kand. khim. nauk.* Moscow, MkhTI, 1985, 109 p.
8.   Kozlov G.V., Shustov G.B., Zaikov G.E. *Zhurn. Prikl. Khimiji,* 2002, v. 75, № 3, p. 485-487.
9.   Argyrakis P., Kopelman R. *Phys. Rev. B,* 1984, v. 29, № 1, p. 511-514.
10.   Smith A., Wilkinson S.J., Reynolds W.N. *J. Mater. Sci.,* 1974, v. 9, № 4, p. 547-550.

THE MULTIFRACTAL ANALYSIS OF DIFFUSION PROCESS IN SEMI-CRYSTALLINE POLYETHYLENE AND ITS MELT

Kozlov G. V.

Abstract

The multifractal nature of the structure and free volume of semi-crystalline polyethylene is proved on the basis of the diffusion properties of this polymer. Such approach allows also to construct the distribution curves of microvoids sizes of free volume. The melt of polyethylene is an Euclidean object with the monodisperse distribution of microvoids sizes. The indicated circumstance defines the different form of dependences of the diffusivity on the molecule size of a gas-diffusant.

Keywords: Multifractal analysis, polyethylene, solid state, melt, free volume, microvoid, gas, diffusion.

INTRODUCTION

Now it is well known [1] that the value of the diffusivity D for polyethylene jumply varies at its melting point T_m. This effect has received the explanation from the positions of the different theoretical concepts. So, in paper [1] this jump is explained from the structural positions, namely, the disappereance of crystallites at T_m. From the positions of a cluster model of the structure of polymer amorphous state the indicated jump is explained by the decomposition of the "frozen" local order in amorphous phase of polyethylene at T_m [2]. And at last, from the positions of the fractal analysis this jump is explained by the transition of polyethylene structure from fractal (solid state) to Euclidean (melt).

However, there is one more fundamental structural aspect of the problem, which, to till now was not given the proper attention. The availability of this aspect implies from two well-known experimental observations: the dependences of the value D for the same polymer on size (diameter) d_m of gas-diffusant molecules [4, 5] and the strong different degree of this dependence for semi-crystalline polyethylene and its melt [1]. In the last case the dependence

$D(d_m)$ is expressed much more weakly than in the first one. These observations allow to make the following supposition. In semi-crystalline polyethylene there is the distribution, typical for polymers, of microvoids size d_v of free volume (where d_v is a diameter of microvoids). Such distribution (or multifractality) of free volume microvoids is a consequence of the multifractality of polymer structure [6]. If to proceed from the approximate condition of diffusion realization $d_m \leq d_v$, it means that in accordance with the increase of d_m in semi-crystalline polymer less microvoids with size d_v, corresponding to the diffusion condition remain, and it levels the differences in diffusion of molecules of gas-diffusant with different d_m. The purpose of the present paper is the check of correctness of the made-above supposition on through semi-crystalline polyethylene and its melt.

EXPERIMENTAL

For the determination of diffusivity D the following equation [1]:

$$D = D_0 \exp\left(-\frac{E_a}{RT}\right), \tag{1}$$

where D_0 is a constant for each gas-diffusant, E_a is the activation energy of diffusion process, R is a universal gas constant, T is the testing temperature.

The values D_0 and E_a for semi-crystalline polyethylene and its melt are accepted under the data of paper [1]. It was supposed that for polyethylene T=293 K, and for its melt T=383 K [1]. The values d_m are accepted under the data of papers [7, 8].

RESULTS AND DISCUSSION

The fractal concept of gas diffusion through polymeric membranes assumes that the diffusivity D can be calculated from the following equation [3]:

$$D = D_0' f_g \left(\frac{l_1}{l_2}\right)^{2(D_f - d_s)/d_s}, \tag{2}$$

where D_0' is a constant, f_g is a fractional free volume, l_1 and l_2 are lower and upper scales of fractal behaviour of the system, accordingly, D_f – a dimension of the localization regions of extra energy, controling processes of gas transport, d_s – a spectral (fracton) dimension.

With reference to the gas diffusion the scales l_1 and l_2 should be defined as follows. The upper scale l_2 is the size (diameter) of a free volume microvoid d_v, and the lower scale l_1 – molecular diameter of a gas-diffusant d_m [3]. From the equation (2) it follows, that under the conditions of D_0'=const, f_g=const and $l_2=d_v$=const the dependence $D(1/d_m)$ in double log-log coordinates should be linear and its slope allows to determine an exponent $2(D_f/d_s)/d_s$. The value d_s owing to its small variation can be accepted as a constant value and equal ~1.45 [3].

In Fig. 1 the dependence $D(1/d_m)$ for diffusion of 9 gases in shown [1] in case of semi-crystalline polyethylene and its melt in double log-log coordinates, where the value D is determined according to the equation (1). As it is possible to see, this dependence sharply differs for two indicated states of polymer: if for semi-crystalline polyethylene the strong decrease D is observed in accordance with the increase d_m, for a melt this dependence is practically absent. As the dependence $D(1/d_m)$ for semi-crystalline polymer is curvilinear, it means the change of the variation of a measurement scale d_m or, with allowance for the condition d_s=const, change of dimension D_f. The value D_f is connected to a fractal dimension of polymer structure d_f by the following relationship [9]:

$$D_f = 1 + \frac{1}{d - d_f},$$
(3)

where d is the dimension of an Euclidean space, in which the fractal is considered (apparently, in our case d=3).

From the equation (3) follows that the change D_f with a scale means the similar change d_f. Approximating the curvilinear dependence $D(1/d_m)$ by the rectilinear segments, from their slope it is possible to calculate the value of the exponent $2(D_f-d_s)/d_s$, then the dimension D_f and from the equation (3) – value d_f for each such part. In Fig. 2 the dependence $d_f(d_m)$ is shown, from which the increase d_f follows in accordance with the increase d_m or the increase of a measurement scale. It is a typical multifractality sing of semi-crystalline polyethylene structure [10]. Let's mark that from the data of Fig. 1 and equation (3), the similar diagram the multifractality of a free volume in semi-crystalline polyethylene, which is a consequence of the multifractality of the structure of this polymer (Fig. 2). Fig. 2 shows the decrease d_f in accordance with the decrease of a scale d_m, i.e., increase of the density of structure with the decrease of a scale [11]. It, in essence, is the definition of a fractal on relationship [9]:

$$\rho \sim R_g^{d_f - d},$$
(4)

where ρ is the density of a fractal object, R_g is its radius of gyration.

As it is known [11], the number of microvoids N_v with the size d_v can be estimated from the relationship:

$$N_v \sim d_v^{-d_f}.$$
(5)

If to assume that N_v corresponds to a relative fraction of microvoids by size d_v, it is possible (after the appropriate normalization of the value N_v) to construct the distribution d_v, shown in Fig. 3. As follows from this Figure, the probability P_v (equal to a relative fraction of microvoids with the size $\geq d_v$) of the microvoid detection with the size which is more than d_v is fast decreased in accordance with the increase d_v. So, the detection probability of a microvoid with $d_v \approx 12$ Å is equal to 0.001.

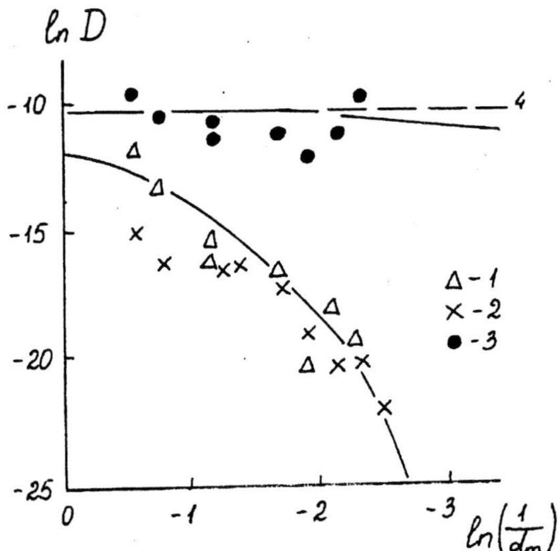

Fig. 1. Dependences of the diffusivity D on reciprocal of molecular diameter of a gas-diffusent d_m in double log-log coordinates for semi-crystalline polyethylene (1,2) and its melt (3,4). Calculation: on the equations (1) (1,3), (2) (2) and (9) (4).

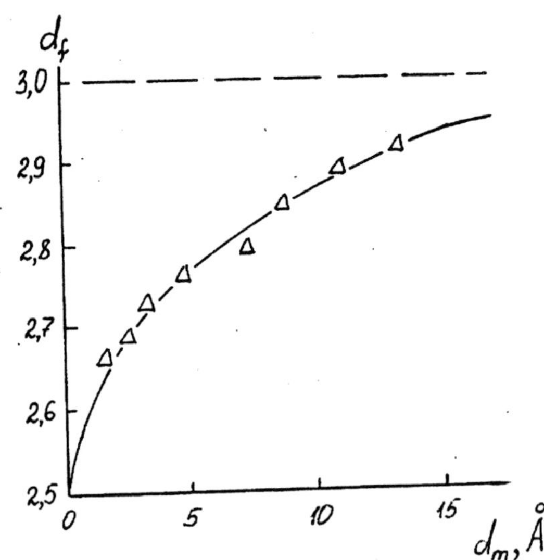

Fig. 2. Dependence of the fractal dimension d_f on a measurement scale d_m for semi-crystalline polyethylene. The dashed line indicats the maximum value $d_f=d=3$.

Now we shall consider the dependence $D(1/d_m)$ for a polyethylene melt, also shown in Fig. 1. At the first approximation its slope is accepted to zero and it means the following condition:

$$2\left(D_f - d_s\right)/d_s = 0. \tag{6}$$

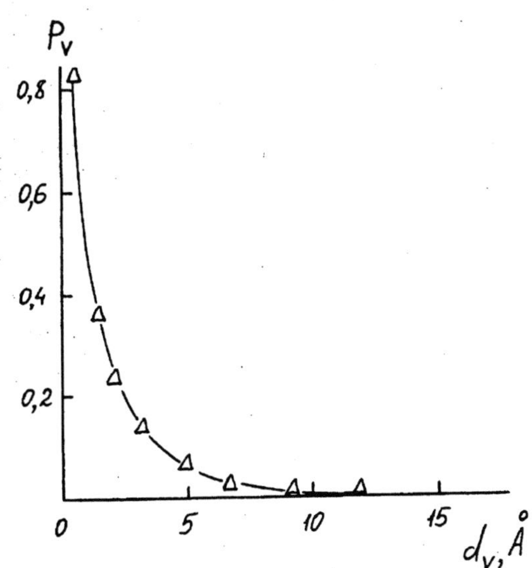

Fig. 3. Dependence of the probability P_v of detection of a microvoid of free volume by the size d_v on the value d_v semi-crystalline polyethylene.

or $D_f = d_s$. For the fractal objects $d_s \leq 2$ [12], and minimum value $d_f = 2$ [11]. From the equation (3) follows, that the minimum value D_f is also equal to 2. It is possible to determine the magnitude d_s from the following equation [12]:

$$d_s = \frac{2d_f}{1+d_f}.$$
(7)

The equation (7) assumes the variation $d_s = 1.33$-1.50 at the greatest possible variation d_f ($2 \leq d_f < 3$ [9]). Thus, it is impossible to receive the condition (6) in the assumption of the fractality of a polyethylene melt. Therefore, it is necessary to assume that the melt is an Euclidean object, in which the dimension D_f loses its sense. From the formal point of view it follows from the equation (3): at $d_f = d$ the value $D_f \to \infty$. From the structural positions it means the following: the dimension D_f characterizes the energetic exitation degree of the loose-packed regions of polymer. In a melt both crystallites, and "frozen" local order are absent, owing to this there is no necessity in the separate characterization of the different structural components and the whole melt is characterized by the dimension $d=3$. For Euclidean objects the identity is correct [13]:

$$d_s = d_f = d = 3.$$
(8)

Therefore the condition (6) for Euclidean objects is trivially fulfilled at the replacement D_f on $d_f = d$.

Therefore, from the above mentioned, the equation (2) in case of a melt becomes:

$$D = D_0' f_g, \tag{9}$$

where the value f_g can be accepted according to Boyer [14] equal $\sim$0.113. Then, for a melt of polyethylene, D up to the constant is equal to D_0' and constant. The value D_0' in this case equal to 2.45×10^{-4} sm^2/s, that is close to the theoretical value D_0^T [1]:

$$D_0^T = \frac{\nu \lambda^2}{6} \approx 10^{-3} \div 10^{-4} \ sm^2/s, \tag{10}$$

calculated from the assumption about a jump of a molecule of a gas-diffusant on the distance λ by some Angstroms with the frequency $\nu = 10^{12}$-10^{13} Hz. Calculated on the equation (9) value D at the mentioned-above values of constants D_0' and f_g is shown in Fig. 1 by the dashed line. The theoretical dependence and it means the absence of the multifractality (distribution) of microvoid sizes of free volume in a melt, that, as it was mentioned-above, is a consequence of losses by the polymer structure of fractal properties.

The dependence of parameter D_0' for semi-crystalline polyethylene calculated on the equation (2) is shown in Fig. 4. As follows from the data of Fig.4, the large values D_0' (about 10^{-5} sm^2/s) only for two inert gases with the least molecule size (He and Ne) are observed, and for remaining gases-diffusants the value D_0' is approximately constant and equal about 10^{-9} sm^2/s. It gives the possibility for the theoretical estimation of diffusivity D for semi-crystalline polyethylene on the equation (2) under the following conditions: $D_0' = 2 \times 10^{-9}$ sm^2/s, $f_g \approx 0.07$, $d_s = 1.45$, $d_v = 12$ Å and the values D_f are accepted with in the interval 4.06-9.78 according to their estimation from slopes of curvilinear dependence D ($1/d_m$), Fig. 1. The comparison of the theory and the experiment shown in Fig. 1, has demonstrated their good correspondence, except for the data for He and Ne for the mentioned-above reasons. Therefore, the correspondence to the theory and experiment for semi-crystalline polyethylene is reached at the inclusion of dependence d_f (or D_f) on a meassurement scale d_m and the inclusion in the scheme of calculation of these scales (d_m and d_v), that uniquely indicates to the multifractality of the structure (and, therefore, of free volume) of this polymer. For obtaining the correspondence to the theory and experiment in case of a polyethylene melt, which is an Euclidean object, the inclusion of the mentioned-above conditions is not required. Let's mark also that the dimension $d_f = 2.67$ corresponds to the probability $P_v \approx 0.84$. It means that the value d_f in the monofractal description of semi-crystalline polyethylene structure should be close to the indicated value d_f. The estimation of the value d_f for polyethylenes by the independent methods confirms this assumption [15].

Further it is possible to construct the multifractal diagram for semi-crystalline polyethylene structure in the terms α-f, where α is a scaling index describing the concentration of singularities, f is a dimension of singularities α [16]. For this purpose the elementary variant is used [17]:

$$P_i \approx l_i^{\alpha_i}, \tag{11}$$

where P_i is the probability of detection of a microvoid with the relative size l_i.

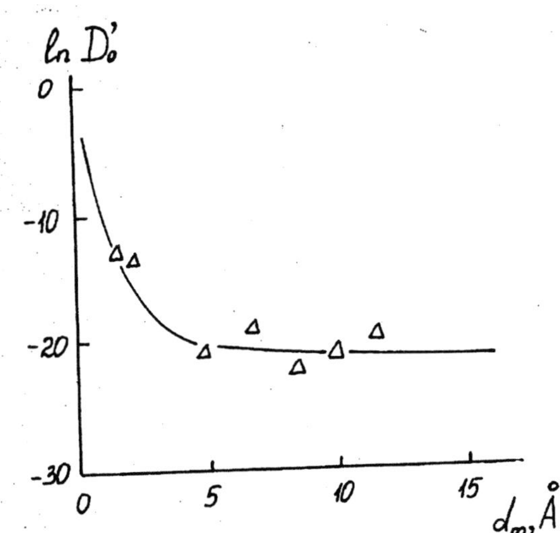

Fig. 4. Logaritmic dependence of a constant D_0' on molecule diameter d_m of gas-diffusant for semi-crystalline polyethylene.

The value P_i is accepted equal to P_v (Fig. 3) and value l_i is determined from the construction of Cantor's set ("dust") as follows. From the data of Fig.3 it follows that the interval of the variation d_v can be accepted equal 0-12 Å. Then the value d_m should be normalized on the value of an interval of the variation d_v, i.e., 12 Å. Further in the construction of Cantor's set from an initial segment of length of 1 its median part (one third) is taken away then the length of two rest parts will be equal ~0.667 [16, 18]. Then this value should be divided proportionally to the size between the surfaces of adjacent microvoids and the size of a microvoid of free volume under the condition $d_v=d_m$ and the latter of the indicated parts corresponds to l_i [18, 19]. The value f is calculated in such a way [18]:

$$f = d_f - 2. \tag{12}$$

The constructed in such a manner diagram α-f for semi-crystalline polyethylene is shown in Fig. 5, where the diagram α-f for a polyethylene melt (trivial case) is also shown, which is straight line with the ordinate $f=1$ ($d_f=d=3$) parallel to axis α. The diagram α-f for semi-crystalline polyethylene has the bell-shaped form typical for such diagrams [6, 10, 16-19]. Its physical sense consists in the following: the curve α-f shows the detection probability in polymer of a microvoid with the size which is more arbitrary d_v at the arbitrary d_f.

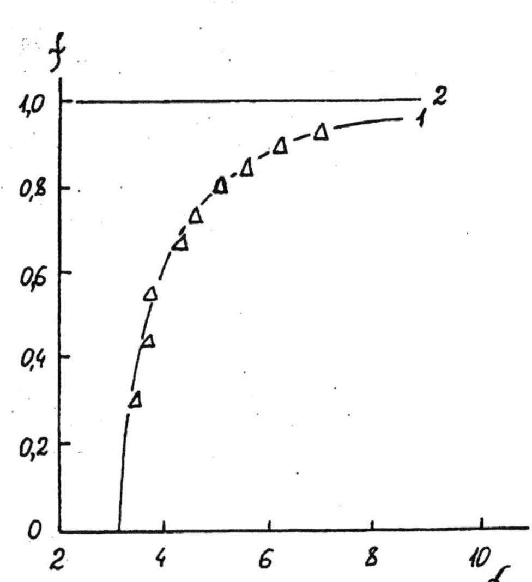

Fig. 5. The multifractal diagrams of the structure of semi-crystalline polyethylene (1) and its melt (2) (trivial case) in coordinates α-f.

CONCLUSIONS

Thus, in the present paper the multifractal nature of the structure and free volume of semi-crystalline polyethylene is proved on the basis of the diffusion properties of this polymer. Such approach allows also to construct the distribution curve of sizes of free volume microvoids. The diagram α-f in usual (already classical) coordinates of the multifractal formalism displays the probability of the existence in polyner the microvoids with the size which is more than their any fixed diameter d_v. The melt of polyethylene is an Euclidean object with the monodisperse distribution of microvoids sizes. The indicated circumstances define the different forms of dependences of the diffusivity D on a size of gas-diffusant molecule.

REFERENCES

1.   Tochin V.A., Shlyakhov R.A., Sapozhnikov D.N. *Vysokomolek. Soed. A*, 1980, v. 22, № 4, p. 752-758.
2.   Afaunov V.V., Mashukov N.I., Kozlov G.V., Sanditov D.S. Izv. Vyssh. Uchenb. Zaved., Sev.-Kav. Region, *Estestv. Nauki*, 1999, № 4 (108), p. 69-71.
3.   Kozlov G.V., Afaunov V.V., Mashukov N.I., Lipatov Yu.S. *Dokl. Nat. Akad. Nauk Ukrainy*, 2000, № 10, p. 140-145.
4.   Reitlinger S.A. *Permeability of polymeric materials*. Moscow, Khimiya, 1974, 272 p.
5.   Rogers C.E. In *Engineering design for plastics* (E. Baer, ed.), Reihold LTD, New York - London, 1965, p. 193-273.
6.   Hayakawa Y., Sato S., Matsushita M. *Phys. Rev. A*, 1987, v. 36, № 4, p. 1963-1966.

7.   Teplyakov V.V., Durgar'yan S.G. *Vysokomolek. Soed. A*, 1984, v. 26, № 7, p. 1498-1505.
8.   Nicolaev N.I. *Diffusion in membranes*. Moscow, Khimiya, 1980, 232 p.
9.   Balankin A.S. *Synergetics of deformable body*. Moscow, MO SSSR, 1991, 404 p.
10. Feder *J. Fractals*. Plenum, New York, 1988, 256 p.
11. Kozlov G.V., Novikov V.U. *Synergetics and fractal analysis of cross-linked polymers*. Moscow, Klassika, 1998, 112 p.
12. Meakin P., Stanley H.E. *Phys. Rev. Lett.*, 1983, v. 51, № 16, p. 1457-1460.
13. Rammal R., Toulouse G. *J. Physiq. Lettr.*, 1983, v. 44, № 1, p. L13-L22.
14. Boyer R.F. *J. Macromol. Sci. – Phys.*, 1980, v. B 18, № 2, p. 461-463.
15. Kozlov G.V., Novikov V.U. *Materialovedenie*, 1999, № 7, p. 22-24.
16. Halsey T.C., Jensen M.H., Kadanoff L.P., Procaccia I., Shraiman B.I. *Phys. Rev. A,* 1986, v. 33, № 2, p. 1141-1151.
17. McCauley J.L. *Intern. J. Modern Phys. B,* 1989, v. 3, № 6, p. 821-852.
18. Williford R.E. *Scr. Metal.,* 1988, v. 22, № 11, p. 1749-1754.
19. Novikov V.U., Kozlov G.V. *Materialovedenie,* 1998, № 10, p. 14-19.

Structure and Kinetic Peculiarities of Thermal Destruction of Some Heterogenic Polymer Blends on the Base of PVC

Kolesov S. V., Kulish E., Chalykh A. E.[*], Aliev A. D.[*], and Zaikov G. E.[**]
Bashkir State University, 32 Frunze Str., 4509074, Ufa, Russia
[*]Institule of Physical Chemistry, RAS, 31 Lenin Prospect, 117915 Moscow, Russia
[**] N.M. Emanuel Institute of Biochemical Physics. RAS, 4 Kosygin Str., 117997 Moscow, Russia

Abstract

The procedure to form a polymeric composition in many ways determines not only the structure of the polymer material hut also the features of its chemical behavior, in particular, in the processes of the thermal destruction. Base on the example of the PVC blends with polyolefins (PE, PP), nitrile butadiene rubbers (CKH-18, CKH-26, CKH-40, polymethylmethacrylate (PMMA), it was shown that at mixing of the components in the solid phase under intense mechanical stresses or at preparation of combined systems from the common solution, the formation of the developed transitional interphase zones of the diffusively mixed components is characteristic. The mixed components can he a non-equilibrium state both on the composition, phase dimensions and on the conformational macromolecules state. It was shown that the formation of the developed transitional interphase layers is accompanied by the significant chain of the PVC in the blend destruction kinetics. The change is expressed in an additional increase of the polymer dehydrochlorination rate. The isothermal heating of samples at $T>T_g$ of the blend components, which provides the acceleration of the phase separation relaxation processes, always leads to the sample thermal stability increase.

INTRODUCTION

In the real conditions of polymeric composites based on the blend of two polymers preparation, the heterogenic systems with a complex enough phase structure and a developed boundary surface are forms. The feature of such systems is their composition non-equilibrium. The high viscosity the low translation mobility of macromolecules at polymer blend formation causes this, and the full phase separation according to the phase diagram [1] does not occur. Because of this, the prehistory of the polymer composition formation determines in many ways both the structure of the polymeric material and properties, in particular, the thermal stability of its individual components. As a good example, we can give the significant increase of the PVC dehydrochlorination rate in its blend with polyolefins (PE, PP) obtained by the combined action of intense stress under pressure with shear (CAIPS) [2]. The increase of the then stability of such composition after the isothermal treatment or after re-precipitation allows to assume t the initially observed decrease of the thermal stability of the halide-containing component is connected with formation of the interphase transitional layers corresponding to the diffusion component blending.

RESULTS AND DISCUSSION

It follows from the electron-microscopy data that such system with the PVC content less than 10 mass % are microheterogenic. Despite the thermodynamic incompatibility of the components, the distinct interphase boundary between the dispersed phase and the media is absent, and some PVC is occluded by the PE phase. From the microphotograph in Fig. 1, it is seen that in the nearby of the PVC particle, there are some oriented macromolecular tangles occurring in the interphase layer between the dispersed phase and the polymeric matrix. The comparison of the oriented tangles zone location with the concentration profile of chlorine a carbon distribution obtained by using the electron-probe microanalysis allows us to assume that PVC macromolecules are incorporated in the tangles along with the polythylene macromolecules. Thus, the most interesting structure-physical change occurring in the blend of two thermodynamically incompatible polymers at CAIPS is formation of the developed transitional layer, non-equilibrium on the composition, phase sizes, conformational state of macromolecules, and so on.

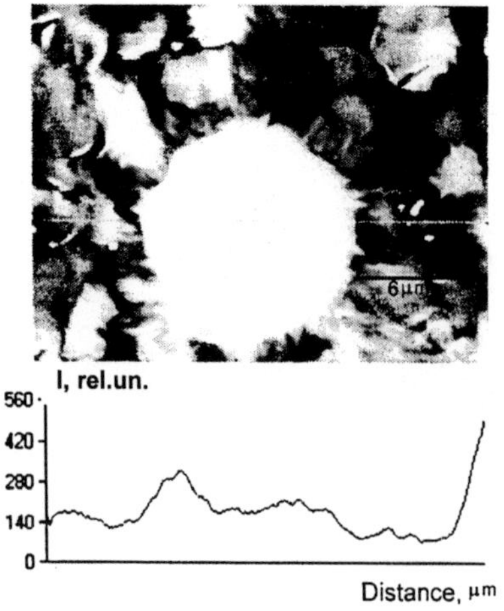

Figure 1. Microphotograph of the PVC particle in the PE matrix in the blend of PVC-PE (90:10 correspondingly) prepared by CAIPS and the concentration profile of chlorine distribution.

This is, probably, the reason of the enhanced (in 5-6 times) PVC decomposition rate in the layer. he analogous heterogenic systems with a diffused interphase boundary are formed from the blends of PVC with nitrile-butadiene rubbers (CKH-18, CKH-26, and CKH-40) or with polymethylmethacrylate (PMMA), which were prepared as thin films from a common solution. The equilibrium phase diagrams for the systems PVC-CKH [3] witness the very low (less than 2-3 mass %) thermodynamic compatibility of polymers, i.e., the inevitability of the heterogenic system formation. However, the level of the heterogeneity depends on the nature of the nitrile rubber and the system forming conditions. For the systems PVC-CKH-18, on the microphotographs (Fig.2), there is a clearly defined heterogeneity. The average phase size is 5-6 μm and the boundaries between phases for the macroscopic CKH-18 inclusions are clearly defined. However, along with the macrophase in the system formed from solution, a microphase is also present. The average size of the microphase particles is 50-60 nm.

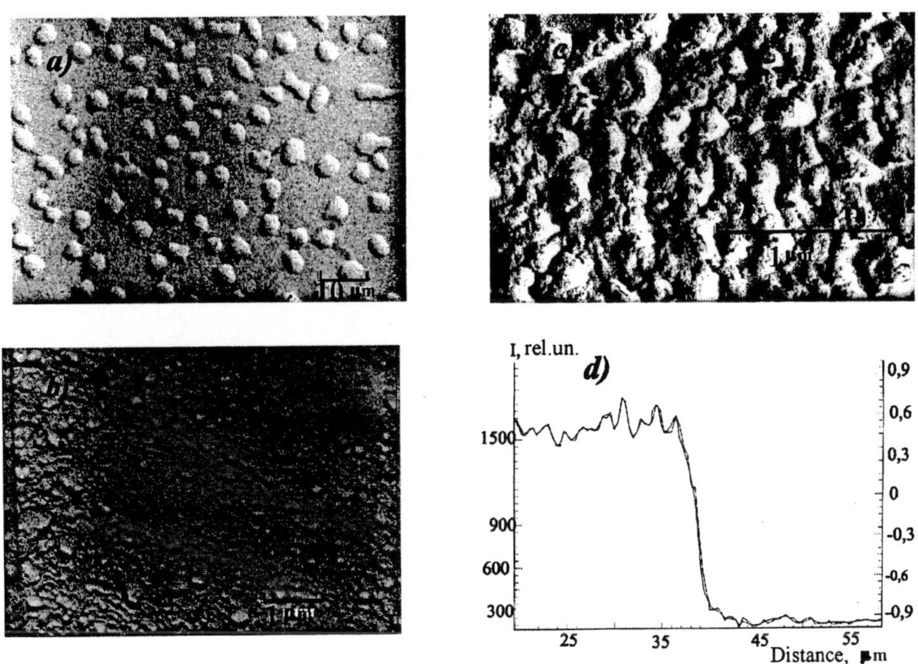

Figure 2. Microphotographs (a-c) and the profile of chlorine distribution (d) for blends of PVC with CKH-18 (a), CKH-26 (b and d) and CKH-40 (c), oobtained from solutions in 1,2-dichlorethane.
CKH=NBR, мкм=μm

The analogous picture is observed also for the system PVC-CKH-26. In this case, the average size of the macroscopic inclusions shifts to the lower values- 0.5-1.0 μm. The system PVC-CKH-40 is related to macroheterogenic systems. In this case the average size of particles of PVC in CKH and CKH in PVC is 10-50 nm. If we assume that such microphase are not independent, then, such system should behave as the single-phase one according to the relaxation characteristics. This is confirmed by the blend glass transition temperatures and by the behavior of the radical-probe (EPR-spectroscopy) in the blend. Hence, in the systems PVC-NBR, there are two types of the phase boundaries: one of them has the sizes comparable with the sizes of the microparticles, the second or separates one macrophase from another one. Based on the results of the electron-probe microanalysis for the NBR macroparticles in the PVC matrix, the clearly defined interphase boundary is absent. On the contrary there is a wide zone corresponding to the diffusion component blending (Fig.2d).

The formation on of the non-equilibrium interphase layer is caused by the retardation of the relaxatic process of the phase separation at film formation, and in its turn, strongly depends on the initial system characteristics-initial solution concentration, and solvent type. This can be seen based on the destruction rate of the PVC-CKH-18 films prepared from solution in 1,2-dichloroethane with different initial concentration (Fig.3a).

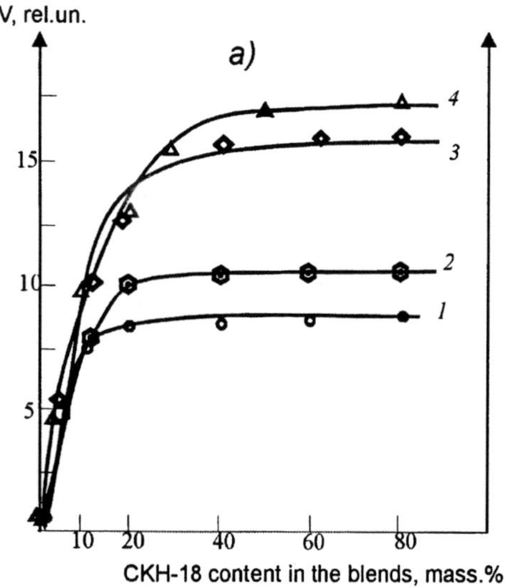

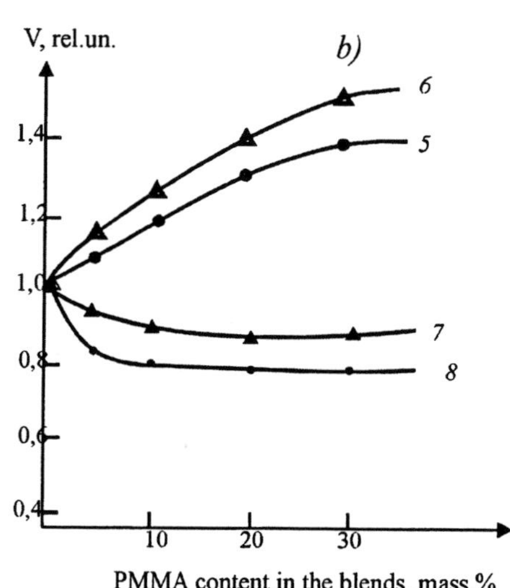

Figure 3. The relative change of the PVC destruction rate in blends with CKH-18 (NBR-18) (a) and PMMA (b) films prepared from solutions in 1,2-dichlotoethane (1,4,7), MEK (5), chlorobenzene (6), and TGF (8) having the initial concentrations (g/dl): 1.0 (1), 2.0 (2), 3.0 (3), 5.0 (4-8).

The effect of the system forming prehistory can be very strong and can fully reverse the character of the Influence of the second polymer on the process of the PVC thermal degradation. The is observed for the PVC-PMMA films prepared from MEK and chlorbenzene (Fig.3b). Just in the case of using a good for both polymers and the complex-forming solvent, the long transition layer without clearly defined boundaries is formed as it is seen based on the chlorine concentration distribution profile . In all of the describe cases, the isothermal treatment of samples at $T>T_g$ of the components leads to the retardation of the system destruction processes what allows to assume that the non-equilibrium macromolecules state in the interphase layers is one of the significant factors for causing the reduction of the thermal stability of PVC in the polymeric blends.

REFERENCES

1.   Chalykh A.E., Sapozhnikova I.N., "Phase equilibrium and interdiffusiou in systems polymer-polymer with chlorine-containing components" (Russ.) // *Uspekhi Chim.* 1984, v. LIII, No.11, pp. 1827-1851.

2. Kolesov S. K, Kulish E.I, Kovarskii A.L, Abalikhina T.M., Mimker K.S. // *Dokl RAS*, 1994, v.3, No.3, pp.335-337.

3. Chalyjh A.E., Gerasimov V.N., Mikhailov Yu.N., *"Diagrams oj the phase state of polymersystems"* (Russ.), M.: Yanus K., 1998.

QUANTUM-CHEMICAL EVALUATION OF RELATIVE ACTIVITY OF ORGANOSILICON METHACRILATES IN RADICAL COPOLYMERIZATION WITH VINYL MONOMERS

N. Lekishvili, T. Guliashvili, G. Lekishvili and L. Asatiani
Javakhishvili Tbilisi State University,3 I Chavchavadze ave.Tbilisi 380028,Georgia

Abstract

The calculation of Quantum-chemical energetic parameters, the distribution of charges on the atom surfaces, the optimal geometry of methacrylic monomers with alkyl(alkoxy)silyl and pentamethyldisioxane groups in the alcohol radical,and also of some widely used vinyl monomers (methylmethacrilate, dimethylvinylacetylenylcarbinol) were performed. Based on the obtained data on the mention above monomer molecules and simplified models of corresponding to them α-radicals, the relative activity of organosilicon methacrilates in a block radical copolymerization with vinyl monomers has been determined.

Key words: Organosilicon methacrilates, evaluation, quantum-chemical, model system, energetic parameters , charges, optimization of geometry

INTRODUCTION

It is known that the copolymerization of two or more monomers opens wide possibilities in synthesis of polymers with given properties. These possibilities are realized both by a choice of certain monomer types and by the regulation of the comonomer links distribution in the macromolecular chain of coplymers [1]. The properties of binary copolymers depend, mainly, on the following three parameters [2]: composition and structure of the macromoleculare chain; distribution of the monomer links in the macrochain; and the compositional inhomogeneity of copolymers. The values of these parameters to a significient extent are determined by the structure and relative reaction ability (activity) of comonomers [1,3]. To evaluate the activity of monomers,the quantum-chemical mthods of calculations are now widely used along with the traditional methods (like kinetic and thermodynamic ones) which are very time labor consuming. The quantum-chemical methods are used to determine the energetic parameters and electronic structure, to optimize the geometry of monomer molecules [4-9]. By comparison of these data and the data of kinetic and thermodynamic parameters (k_p, k_o, V_{orig}, V_{max}, r_i, r_j, Q, e, ΔH_{sum}, etc.}, it is possible to evaluate fully enough the relative activity of monomers and, therefore the compositional inhomogeneity of copolymers both on the composition, and on type of monomer links distribution in macrochains of copolymers[1,3].

From the point of view, the preliminary evaluation of the relative activity of novel and known before unsaturated monomers in radical copolymerization with vinyl monomers by using the quantum-chemical calculation can give a valuable information for quantitive description of the reaction. These calculations can be also used to discuss the"behavior"of all the comparable monomers in the process when the same method of calculations is used [10,11].

RESULTS AND DISCUSSION

The quantum-chemical calculation of energetic parameters and electronic structure of molecules of organosilicon methacrylates (SiMMA) [10,12,13] were conducted by using the semi-empirical method MO LCAO SCF in approximation "PM-3" with the standard geometrical parameters [14]. Because in the literature there are no data on the special structures of the mentioned above monomers, the optimization of geometry of the geometry of the monomer molecules based on the some parameters was performed by using the method of molecular mechanics "MM-2"[15]. For comparison, the quantum-chemical parameters of methylmethacrylate(MMA) and dimethylvinylacetylenylcarbinol (DMVAC) were calculated (Fig.1,and Table 1).

The basic reaction center in reactions of radical copolymerisation is vinyl group. Therefore , the values of calculated energetic parameters and electronic structure are given only for this group. From the data in Fig.1 and Table 1, it is seen that for monomers MMA and SiMMA(1-3)

There are approximately the same charge parameters of the carbon atoms of the vinyl group. Close values of these parameters have also the other organosilicon methacrylates (M_5 and M_6). However, they difer in a certain degree from the corresponding parameters of the MMA. The quite different charges distribution takes place on the C_α and C_β atoms in the molecules of a monomer having less-conjugated (vinylacetylene) chain of the dimethylvinylacetylenylcarbinol (DMVAC) Fig.1,Table 1) [10]. Some certain dependencies were established also for the calculated characteristics of the vinyl group: $P^\pi_{C=C}$, : $P^{\sigma+\pi}_{C=C}$, the index of free valency, $F_{C\beta}$, and also the geometric(μ_{SUM}, μ_X) and energetic (E_{HOMO}, E_{LUMO}, and ΔH_{form}) parameters on the structure of the methacrilic monomers (Table 1). This allows to determine their different ability to radical (co)polymerization.

1).
-0,0710
-0,0627
-0,1613
-0,2434
+0,4064
-0,3988
A
-0,0718
-0,0755
-0,1288
-0,2610
+0,4625
-0,5222
B
2).
-0,0787
-0,0627
-0,1599
-0,2160
R1
+0,3896
-0,4063
A
-0,0879
-0,0748
-0,1239
-0,2031
R1
+0,4326
-0,5664
B
3).
-0,0737
-0,0622
-0,1604
-0,2459
R2
+0,4093
-0,4017
A
-0,0677
-0,0637
-0,1212
-0,2463
R2
+0,4142
-0,6064
B
4).
-0,0746
-0,0627
-0,1598
-0,2447
R3
+0,4095
-0,4025
A
3
-0,1010
-0.1013
-0,1619
-0,2476
R3
+0,4787
-0,4897
B

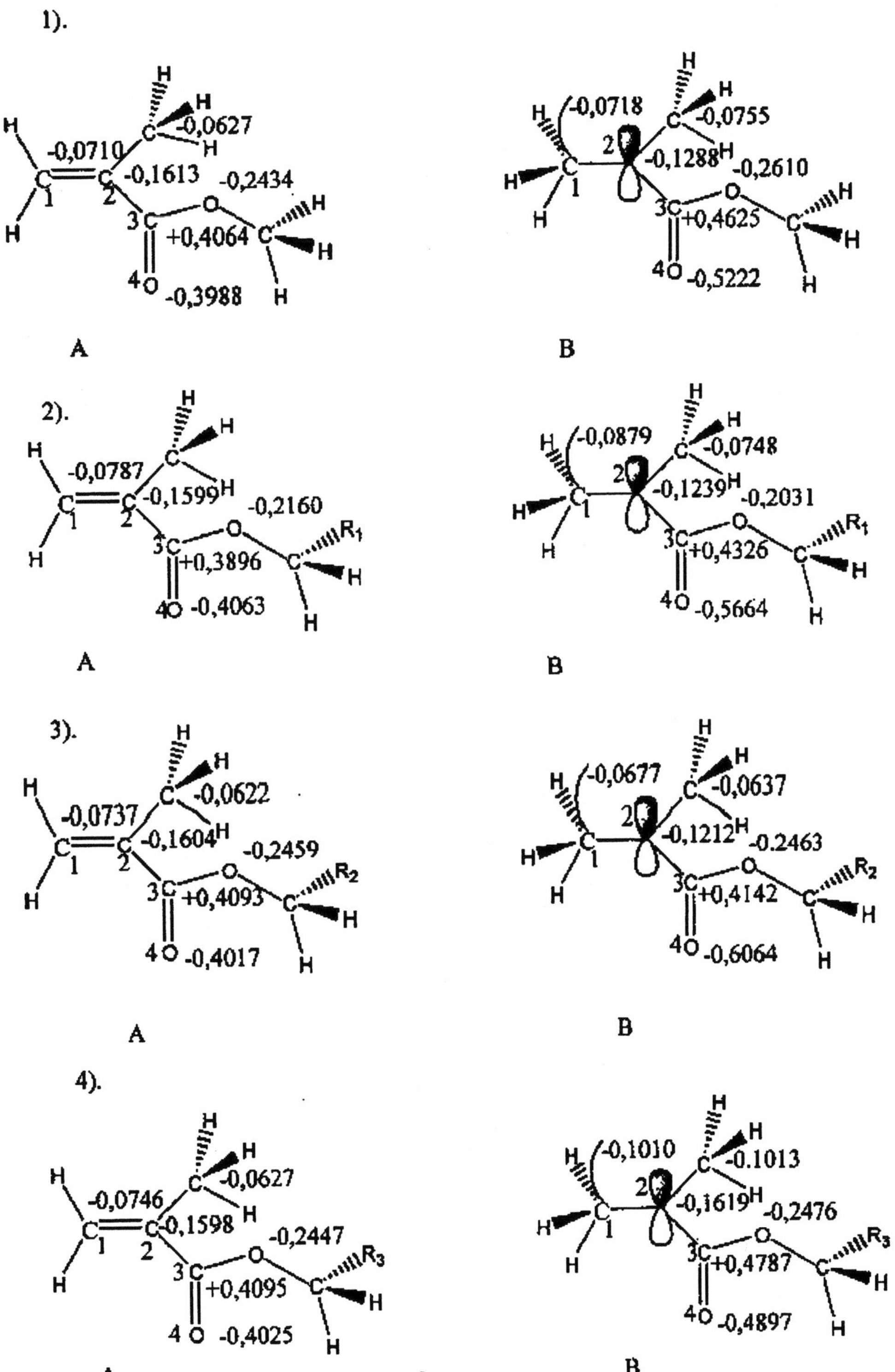

Figure 1.

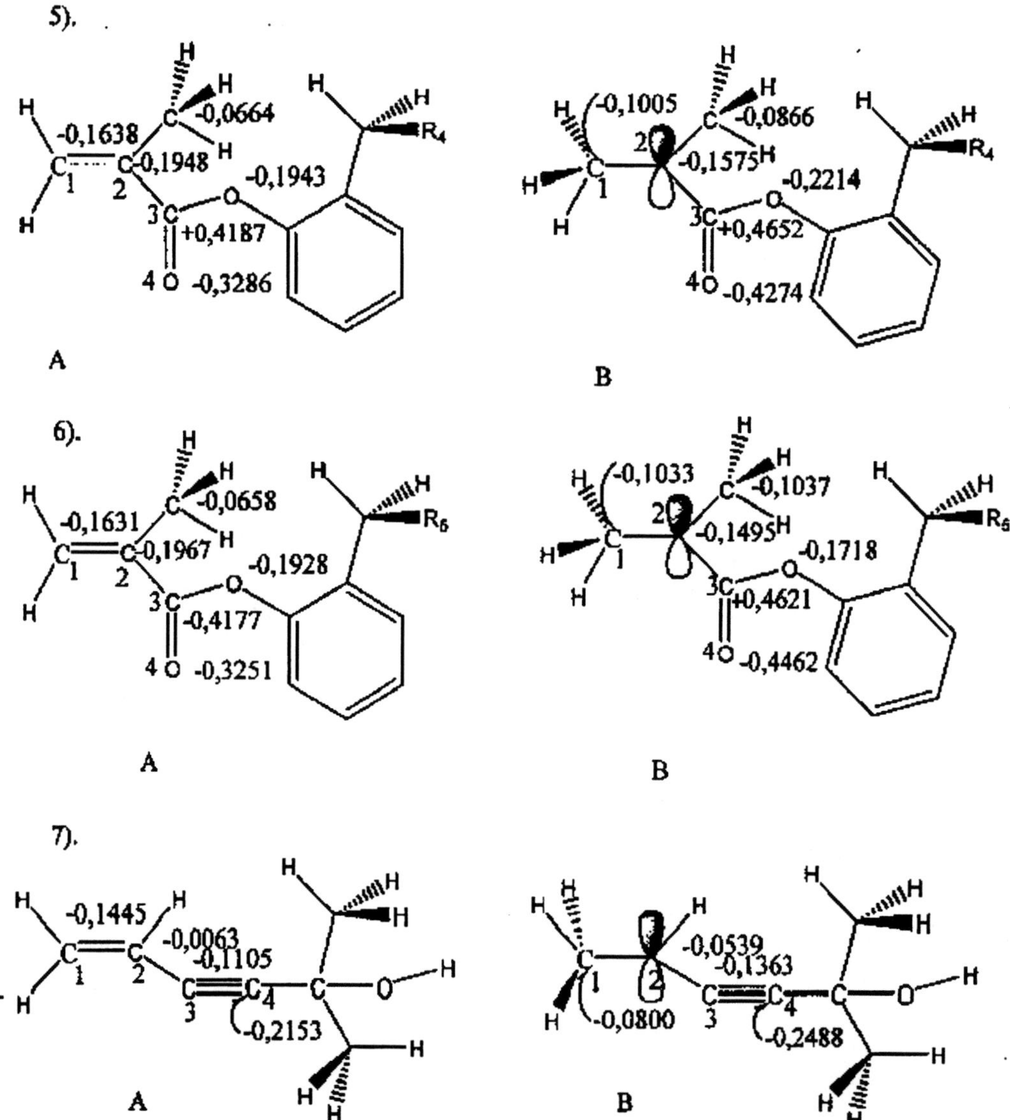

Figure 1. Spacial structure and distribution of charges on the atoms of the vinyl group of monomers (A) and corresponding to them α-radicals (plane structure) (B).
1.- MMA-M_1; 2.-Methacryloxymethylpentamethyldisiloxane; R_1=Si(CH$_3$)$_2$OSi(CH$_3$)$_3$ -M_2 ; 3.-Methacryloxybutylpentamethyldisiloxane: R_2= (CH$_2$)$_3$ Si(CH$_3$)$_2$OSi(CH$_3$)$_3$-M_3 ; 4.-Methacryloxypropyldiethoxymethylsilane: R=(CH$_2$)$_3$Si(CH$_3$)$_2$(OC$_2$H$_5$)$_2$,-M_4; 5.1-Methacryloxyphyl-6-propylpentamethyldisiloxane :R_5= Si(CH$_3$)$_2$OSi(CH$_3$)$_3$-M_5; 6.-1-Methacryloxyphyl-6-propylmethyldiethoxysilane ; R_6=(CH$_3$)$_2$(OC$_2$H$_5$)$_2$,-M_5; 7.-dimethylvinylacetylenylcarbinol (DMVAC)-M_6.

Table 1. Calculated charges and energetic parameters of molecules of siliconcontaning methacrilates, methylmethacrilate, and dimethylvinylacetylenilcarbinol

M_i	q_α	q_β	$P^\pi_{C=C}$	$P^{\sigma+\pi}_{C=C}$	$F_{C\alpha\beta}$	ΔH_{form} kDj/mol	E_{HOMO}	E_{LUMO}	$\mu_{SUM,}$ Deb.	$\mu_{X,}$ Deb.
M_1	-0.0710	-0.1613	0.9385	1.9084	0.7869	298,97	-10.410	-0.03	1.72	0.70
M_2	-0.0787	-0.1599	0.9413	1.9114	0.7869	949.89	-9.138	+0.10	2.57	1.71
M_3	-0.0737	-0.1604	0.9395	1.9095	0.7863	904.10	-9.073	+0.06	2.28	1.63
M_4	-0.0743	-0.1600	0.9396	1.9095	0.7862	1040.11	-9.523	+0.09	2.60	3.13
M_5	-0.0638	-0.1948	0.9332	1.9034	0.7870	858.29	-8.999	-0.29	2.76	0.69
M_6	-0.6310	-0.1967	0.9330	1.9034	0.7879	882.36	-9.200	-0.30	3.58	1.35
M_7	-0.1445	-0.0063	0.9503	1.9221	0.7765	37.90	-9.799	+0.204	1.67	1.37

The electronic structure of methacrilates plays significant role in ability of monomers (Mi) to interact (during the process) with the growing radicals ($R^\bullet_m$) (with a probable change of the hybrid state of the corresponding carbon center)[10]: $M_i + R^\bullet_m \rightarrow R_m - M^\bullet_1$

For the purpose of the current investigations, the specific interest is the energetic state and structure of the free radicals modeling the end chain of the growing macromolecular chains from the vinyl monomers of the general type $-CH_2-C^\bullet HX$ [10]. The corresponding simplest models are α-radicals of the type $CH_3-C^\bullet HX$. The choice of these radicals is based on the calculated data according to which for the all investigated monomers, the energetically preferable action is addition of monohydrogen to the β-carbon atom of the vinyl group with localization of the non-pared electron mainly on the α-carbon center (depending on the X nature) [11].

Due to absence in the literature the data on electronic structure and geometry of α-radicals corresponding to the investigating monomers, we performed calculation of their electronic structure and geometry optimization in the same way as for monomers, i.e. by using "PM-3"and the "MM-2" methods correspondingly. These methods, compared with other methods of calculation, are less time consuming and allow to obtain better results for such model systems [14].(Fig.1, A and B, Tables 2a,b and 3).

From the calculated data (Tables 2a,b and Fig.1, B), it is seen that the optimization of the geometry of the growing α-radicals from the methacrilic monomers by using the molecular mechanics method MM-2 predicts the planetrigonal structure of these radicals C_α atom of the double bond of the vinyl group.On the atom , the non-paired electron is concentrated. The value of the spin density , $\rho_{C\alpha}$,depends on the alcohol radical type and is slightly different from the corresponding value of C_α in DMVAC(Table 2a,b).

The energetic parameters of the model α-radicals have been also calculated (Tables 2a,b and 3). These parameters along with the energetic parameters of the monomer allow to a certain approximation to evaluate the probability of energetical passage of an electron from the HOMO α-radical onto the LUMO monomer [16] at attachment of the monomer to its own radical or the "foreign" radical. In these cases, naturally, it is very difficult but necessary to take into account in the real systems the spacial hindrances at the α-carbon atom of the growing radical(what requires an additional information)[10].

By comparing the calculated energetic parameters of the model α-radical with corresponding data for a monomer (Table 3), it is possible to make a conclusion that the lower the E_{LUMO} of the monomer and also the narrower the so called "energetic split" between E_{LUMO} of monomer and E_{HOMO} of the α-radical of corresponding to it monomer, the higher should be the reaction ability of the monomer [16].

Taking into account the above considerations, the investigating monomers based on their relative activity can be arranged in the following row: $M_5 \cong M_6 > M_1 \cong M_2 \cong M_3 \cong M_4 > M_7$. Based on the conducted quantum-chemical calculation , it can be assumed that the most successful pairs for the obtaining at blok radical copolymerization compositionally inhomogeneous copolymers both on composition and on the type of monomer link distribution in the macromolecular chain, are the organosilicon methacrilates - DMVAC (Fig.1) [17,18]. However, the processes of complex formation [13] and diffusion, media viscosity, etc. should be taken into account at formation of the real systems [1].

CONCLUSIONS

The quantum-chemical calculation of the energetic parameters electronic structure, and optimization of the geometry of some organosilicon methacrylates having different structure of the alcohol alkyl radical (containing Si atoms), methylmethacrilate and dimethylvinylacetylenyl-carbinole,and also corresponding to them α-radicals (by using the model systems of the"simpli- fied" α-radical of the $CH_3\text{-}C^{\bullet}HX$ type) have been performed.Based on the calculated parameters, the preliminary evaluation of the relative activity of the organosilicon methacrylates in radical copolymerization with MMA and dimethylvynilacetylenylcarbinole has been studied. It was established that the copolymers having better compositional homogeneity of the macro chains both on composition and on the character of monomer links distribution can be obtained based on the methylmethacrylates and the organosilicon methacrila-tes with the disiloxane groups in the alcohol alkyl radical.

Table 2a. Calculated charge and energetic characteristics of simplified models of α-radicals of organosilicon methacrylates, methylmethacrylate, and dimethylvinylacetylenylcarbinole

Radicas and their structures (M_i)	P^{π}_{1-2}	$\rho_{C\alpha sum}$	$\rho_{C\alpha z}$	$\rho_{Carb.}$	$\Delta H_{form.}$, kDj/mol	E_{SUMO}	E_{LUMO}	$\mu_{SUM,}$ Deb.	$\mu_{x,}$ Deb.
M_1									
Plane	0.0632	0.7272	0.7272	0.0780	-319.49	-5.0640	+1.2758	3.397	0.012
Pyramid.	0.0576	0.7885	0.5243	0.0293	-296.27	-5.4877	+0.6400	1.613	1.270
M_2									
Plane	0.0662	0.7109	0.6868	0.0841	-1000.18	-5.0255	+0.8240	4.257	1.028
Pyramid.	0.0575	0.7414	0.5373	0.0497	-963.70	-5.3327	+0.8849	4.643	1.705
M_3									
Plane	0.0627	0.7323	0.7323	0.0754	-1055.41	-4.9862	+1.2258	4.067	0.672
Pyramid	0.0578	0.7378	0.5261	0.0545	-1024.93	-5.4592	+1.0495	-	-
M_4									
Plane	0.0642	0.7291	0.7291	0.0763	-1072.17	-5.0055	+1.0200	5.981	1.434
Pyramid	0.0573	0.8001	0.5247	0.0159	-1046.45	-5.4322	+0.6445	3.872	3.716
M_5									
Plane	0.0669	0.6811	0.6807	0.1091	-879.48	-5.2578	+0.0189	5.778	0.808
Pyramid	0.0603	0.7433	0.5580	0.0585	-855.07	-5.4503	-0.0463	4.563	0.732
M_6									
Plane	0.0693	0.6090	0.6090	0.1541	-958.76	-5.5567	-0.1177	8.749	2.073
Pyramid	0.0579	0.7870	0.5295	0.0293	-924.44	-5.5474	-0.1758	3.766	0.472
M_7									
Plane	-	0.7186	0.7186	0.2094	34.09	-4.6962	+1.1769	2.736	+0.18
Pyramid	0.0537	0.7197	0.4934	0.1722	73.90	-5.3193	+1.1746	2.738	1.441

REFERENCES

1. Ivanchev S.S., *Radical polymerization* (Russ), Leningrad, Chimia., 1984,p.144.

2. Kuchanov S.I., Modern aspects in quantitative theory of radical copolymerization.In book: *Reaction in polymeric systems* (Russ), Ed. S.S. Ivanchev. Leningrad,Chimia, 1987. Chapter 4, pp.109-110.

3. Lekishvili N.G., RubinstainG.M., and Asatiani L.P. / *Polymeric light conductors: formation, properties, perspectives*. Publ. Tbilisi Université, 1994, 131 p.

4. Lekishvili N., Guliashvili T., Asatiani L.,et. al. // *Intern.J.Polym. Mater.*, 1995, v.27, pp. 163-174.

5. Lekishvili N., Zaikov G., Lachinov M. // *Polymer Yearbook*, 2002 (In press).

6. Lachinov M.B., Guliashvili T.T., Lekishvili N.G. et. al.// *Russian Polymer News*, 1999, v.39, pp.27-33.

7. Bayoras G., et.al. // *Visokomolek. Soed.*(Russ.), 1987, v. A29, No4, p.685.

8. Andrianov K.A., Volkova L.M., Zhdanov. // *Abstracts of the All-Union (URSS) Conference on chemistry and application of organosilicon compounds* (Russ). Jenuary, 1980, GNICHTEOS,p.223.

9. Lekishvili N., Asatiani L.,et.al. // *Bull.of The Georgian Acad.of Sci.*(Russ), 1994, v.152, No3, p234..

Table 2b. Calculated charge and value of index of free valence of simplified models of α- radicals of organosilicon methacrylates, methylmethacrylate, anddimethylvinylacetylenylcarbinole

Radicals and their struct. (M_i)	$P^{\sigma+\pi}_{C1-C2}$	P^{σ}_{C2-C4}	$P^{\sigma+\pi}_{C2-C4}$	$F_{C\beta}$(radical)- $F_{C\beta}$(Monom.)	F_α	$F_{C\beta}$
M_1						
Plane	1.0332	0.1444	1.0452	0.0201	0.8164	0.7668
Pyramid.	1.0175	0.0512	1.9372		0.8866	0.7652
M_2						
Plane	1.0330	0.1529	1.0588	0.0202	0.8110	0.7667
Pyramid.	1.0162	0.0931	0.9749		0.8849	0.7644
M_3						
Plane	1.0238	0.1409	1.0337	0.0203	0.8173	0.7650
Pyramid	1.0152	0.0964	0.9192		0.8852	0.7648
M_4						
Plane	1.0249	0.1433	1.0369	-	-	-
Pyramid	1.0080	0.0407	0.9184		-	-
M_5						
Plane	1.0265	0.0670	1.0273	0.0210	0.8217	0.7666
Pyramid	1.0159	0.0603	1.0129		0.8773	0.7649
M_6						
Plane	1,0316	0.0688	1.0328	0.0209	0.8147	0.7670
Pyramid	1.0086	0.0581	1.0071		0.8877	0.7665
M_7						
Plane	-	1.7077- C_3-C_4 ;	2.6983 C_3-C_4 ;	-	0.8217	-
Pyramid	1.0209	1.7436 C_3-C_4	2.7338 C_3-C_4		0.9136	0.7607

10. Lekishvili N.,Andguladze N.,et.al. // *Bull.of The Georgian Acad. of Sci.*(Russ),1995, v.152, No3, pp.537-541.

11. Aisner Yu.E. ,and B.I.Erusalimsky. *Electronic aspects of polymerization reaction* (Russ). Leningrad, Nauka. 1976. 179 p.

12. Marker R.L.,and L.E.Note // *J.Org. Chem.* 1956,V.21,No 12,pp. 1337-1339.

13. Andrianov K.A., Volkova L.M., et.al. *Zhurnal Obshei Chimii* (Russ), 1980, L(CXII), p.1088.

14. Stewart J.J.P. // *J.Comp. Chem.*, 1989,v.10, p.209.

15. Clark T., *Comput. Chemistry* (Russ),. Moscow, Mir, 1989(Transl. from English).

16. 16.Asatrian R.S., Mailian N.Sh., Kharatian Y.G., and G.V. Asatrian // *Visokomolek. Soed.*(Russ.), v.33B,

17. *Short communic.*, 1991, No2, pp.91-94.

18. Lekishvili N., Asatiani L., Andguladze N., and T.Guliashvili. *Intern.Symposium on Organosilicon Chemistry.* 1-6, Sept., Montpelier, 1996,PB-64.

19. Lekishvili N., Asatiani L., Guliashvili T., and G. Lekishvili. *"Andrianov's Reading"*. Preprints. Moscow, 17-19 Jan., 1995, p.44.

Carbo-Functional Oligoorgano-Siloxanes: Structure, Reaction Ability and Property

D. U. Murachashvili*, L. M. Khananashvili*, and V. M. Kopylov**

*I.Javakhishvili Tbilisi State University,3 Chavchavadze ave, Tbilisi 380008,Georgia
** GNIIKHTEOS, 38 Highway Entusiastov, Moscow111123, Russia

Abstract

The interaction between epoxyorganosiloxanes and amines has been investigated. it was shown that the use of aminohydroxysiloxanes in rubber compounds significantly improves some of their physical properties: modulus, tensile, fatigue, resistance to heat aging. The use of these compounds increases the hydrophobicity of commercial building floors coatings and their resistance to different synthetic

Keywords: tpoxy organosiloxane, aminohydroxysiloxane, parallel-consecutive two-steps reaction, cooligomerization, kinetics.

Introduction

Carbo-functional organosilicon compounds can be used as original raw materials in synthesis of polymers with organo-inorganic molecular chains. In particular, the amino-hydroxysiloxanes (AHOS) are of special interest because they are key compounds for preparation of carbofunctional organosilicon oilgomers and polymers [1,2]. One of the reactions leading to formation of (AHOS) is the reaction of epoxy organosiloxanes with amines [3-6]. The AHOS can be formed by co-oligomerization of organocyclosiloxanes with silicon containing dimer amino- alcohol in presence of nucleophilic initiators [7,8].

We have investigated the interaction of (3- glicyloxypropyl) tetramethyldisiloxane(I) [9] and 1,3-bis(3-glicydoxypropyl)tetramethyldisiloxane (II) [10] with diethyl amine, bis[(dimethylbutylamino)ethylamine, allylamine and aniline.

The co-oligomerization of obtained AHOS with oktametylcyclotetrasiloxane (D_4) in presence of tetramethyl ammonia siloxanolate (STMA), and also with addition of dimethylformamide (DMFA). The synthesized AHOS were used as modifiers of rubber compounds and as solidifiers of epoxy resins.

EXPERIMENTAL PART

The IR-spectra were obtained by using the "SPECORD" spectrometer in the range 4.000 - 400 sm^{-1} for samples solutions in CCI$_4$. The NMR -^{1}H and ^{13}C were recorded on the "Bruker" spectrometer AM-360. The chromatography analysis of original reagents and the .reaction products were performed by using the device LKhM-80, type 2 (the column 3000 x 4 mm, the head -"Chromosorb W, the phase-5 mass % SE-30, gas-carrier-helium). The quantitative determination of functional groups were performed by using procedures described in [11,12].

Synthesis of aminohydroxysiloxanes. The compound 1 was placed in a quadruple neck flask at room temperature. The calculated amount of amine in alcohol solution was added to the flask drop-by-drop. The mixture was heated to 45-50°C and the synthesis was continued till completion of the reaction. The solvent an the unreacted compounds were removed by vacuumizing at 60°C and residual pressure of 266 Pa. The characteristics of the obtained products are given below.

a).   (3-Diethylaminopropanol-2- oxypropyl)tetramethyldisiloxane (Ill). From 2.48g (0.01 mol) of compound I and 1.46g (0.02 mol) of diethylamine it was synthesized 3.14 g (98%) of compound III, n$_D$^{20}=1.4395.

b).   (3-Allylaminopropanol-2-oxypropyl)tetramethyldisiloxane (IV). From 2.48g (0.01 mol) of compound I and 2.28g (0.04 mol) of allylamine it was synthesized 2.92g (96%) of compound IV, n$_D$^{20}=1.4412.

c).   Bis[(propanol-2-oxypropyl)tetramethyldisiloxane]allylamine(V).   From   4.96g(0.02 mol) of compound I and 0.57g (0.01 mol) of allylamine it was synthesized 5.36g (97%) of compound V, n$_D$^{20}=1.4478.

d).   (3-Anilinopropanol-2-oxypropyl)tetramethyldisiloxan (VI). From 2.48g (0.01 mol) of compound I and 0.93g (0.01 mol) of aniline it was synthesized 3-30g (97%) of compound VI, n$_D$^{20}=1.4785.

e).   Bis(propanol-2-oxypropyl)tetramethyldisiloxane)aniline(VII). From 4.96g (0.02 mol) of compound I and 0.93g (0.01 mol) of aniline it was synthesized 5.77g (98%) of compound VII, n$_D$^{20}=1.482i.

f).   1,3-Bis[3-(2-hydroxy-3-allylaminopropoxy)propyl]tetramethyldisiloxane(VIII); 1,3-Bis [3-(2-hydroxy-3-diethylaminopropoxy)ropyl]tetramethyldisiloxane (lX) and 1,3-Bis {3[-2-hydroxy-3-(dimethylbuthylaminoethylamine)propoxy]propyl}tetramethyl disiloxane (X) were prepared according to procedure described in [4].

The elemental composition and MM of compounds I-X correlated with the calculated ones.

The co-oligomerizatlon of D$_4$ with compounds VIII, IX and X correspondingly were performed according to procedure described in [7]. The α, ω-Bis[3-(2hydroxy-3-proroxy)propyl]oligodimethylsiloxanes - XI, XII and XII- correspondingly with number of Si atoms of 40 were synthesized.

Rubber Compound Preparation

Rubber compounds based on non-polar unsaturated elastomers of general type were prepared in an internal mixer (volume 2 liters) in two stages. The discharging temperature of the first stage was 145-150°C. Mixing time was 4 min. Compounds VIII, IX, X, XI and XII were incorporated into rubber mix together with curing agents on roll-mill at 60-65°C for 8 min. The sample curing was done in a steam press at 143±1 °C for 45 min. The sample tests were performed at room temperature both original samples and samples after heat aging at 100°C for 96 hrs.

Coating for Floors of Commercial Buildings

The coating was prepared in the following way: a block-copolymer of ethylene oxide and propylene was added to water, the mixture was stirred until copolymer was dissolved completely. Then, the epoxydiene resin, ED-20 was added to the obtained solution. The mixture was dispersed by stirring at 1000-1500 rev./min for 15-20 min. To the obtained emulsion the compounds VIII, X or XII were added immediately before the acoating application.

RESULTS AND DISCUSSJON

It was found that the interaction of compound I with diethyl amine occurs mainly according to reaction with the secondary alcohol group:

$$(C_2H_5)_2NH + \underset{\underset{O}{\diagdown\diagup}}{CH_2\text{-}CHCH_2}O(CH_2)_3Si(CH_3)_2OSi(CH_3)_2H \rightarrow$$

$$\overset{1 \quad\quad 2 \quad 3 \ 4 \ 5 \ \ 6 \ 7 \quad\quad\quad 8 \ 9}{HSi(CH_3)_2OSi(CH_3)_2CH_2CH_2CH_2OCH_2CH(OH)CH_2N(H)CH_2CH_3}$$

$$III$$

The composition III has been confirmed by existence of signals in ^{13}C NMR spectra related to carbon atoms C(6) - C(8) with δ 47.5, 67.4 and 74.6 m.d. correspondingly.

The reaction of compound I with allylamine was performed at molar ratio of original regents 1:1, 1:4 and 2:1. It was found that in the reaction products were compounds both mono-addition(IV) and di- addition (V):

$I + CH_2=CHCH_2NH_2 \rightarrow$

$$\underset{1\quad\quad\quad 2\ 3\quad\ 4\quad\ 5\quad\ 6\quad\quad 7\quad 8\quad\quad\quad 9\quad\quad\quad\quad\ 11\ 12}{HSi(CH_3)_2OSi(CH_3)_2CH_2CH_2CH_2OCH_2CH(OH)CH_2N(H)CH_2CH=CH_2}$$
$$IV$$

$$[HSi(CH_3)_2OSi(CH_3)_2(CH_2)_3OCH_2CH(OH)CH_2]_2NCH_2CH=CH_2 \tag{2}$$
$$V$$

(Here and further in the text the numbering of the carbon atoms in mono- and di-addition products is the same).

In the ^{13}C NMR spectra of compounds IV and V there are signals corresponding to C(8) - C(12) what is witnessing the reaction occurring to the scheme (2).

The addition reaction of I to aniline was performed at molar ratios of original reagents 1:1 and 2:1:

$$I + PhNH_2 \rightarrow$$
$$\rightarrow HSi(CH_3)_2OSi(CH_3)_2(CH_2)_3OCH_2CH(OH)CH_2N(H)Ph +$$
$$VI$$
$$[HSi(CH_3)_2OSi(CH_3)_2(CH_2)_3OCH_2CH(OH)CH_2]_2NPh \tag{3}$$
$$VII$$

The ^{13}C NMR spectra of compounds VI and VII show the appearance of the (7) and (8) signals what confirmed the suggested structures.

Investigation of the reaction between I and allyl amine and aniline with GPC showed that the difference in reaction activity of the hydrogen atoms of the aniline amine group is significantly higher than in allyl amine. So, at equimolar amounts of original reagents the ratio of mono- and di-addition in the reaction products for aniline and allyl amine is 90:10 and 65:35 correspondingly.

In a general view, the reaction between aniline or allyl amine and compound I can he descrihed as a two-step process occurring according to Eqs (4) and (5) and the reaction with diethyl amine with compound I occurs as a one-step process according to Eq. (4)

$$D + I \xrightarrow{\ k_1\ } A_1 \tag{4}$$

$$A_1 + I \xrightarrow{\ k_2\ } A_2 \tag{5}$$

where: D -amine, A_1 and A_2 - the reaction products. The values of k_1 and k_2 are given in Table I.

Table 1. Rate constants k_1 and k_2 of reactions of compound I with amines.(Calculations were performed according to [13]).

D	I:D	$[A_1]$	$[A_2]$	I_o	D_o	k_1 $\times 10^{-1}$ I/(mol x sec)	k_2
AII NH₂	1:4	93	7	0.4	1.68	12.2	3.63
	1:1	65	35	0.7	0.66	12.2	3.63
	2:1	5	95	0.7	0.37	12.2	3.63
Et₂NH	1:2	98	-	0.5	1.05	7.68	-
PhNH₂	1:1	90	10	0.6	0.60	11.0	0.30
	2:1	2	98	0.7	0.35	11.0	0.30

Comparison of the rate constant of the first step shows that they are close with a small variation. Based on the one, the rate constant can be arrange in the following row by decreasing: $AIINH_2 > PhNH_2 > (Et)_2NH$.

For the parallel-consecutive two-steps reaction [13], the ratio of the first to second step process rate constants, x was calculated. From the ratio $x = k_1/k_2$ the values of constant k_2 were calculated and presented in Table 1. It was found that in case of aniline the value of k_2 is significantly lower compared with allyl amine.

In the IR spectra of compounds III-VII there is no absorbency bands of 3060 and 915sm^{-1} corresponding to epoxy group. However, an intensive peak at 3400 sm^{-1} appears. This band corresponds to OH/NH groups. It was shown earlier, that during co-oligomerization of organocyclosiloxanes with carbo-functional disiloxanes in presence of DMFA the rate of siloxane ring opening is increasing and the content of low molecular products and linear short-chain oligomer reduces at achieving the approximate constant value of η_{in} [7,14,15]. Because of this, in this work the co-oligomerization of D_4 with compounds VIII-X was conducted with addition of DMFA. The synthesis way can be described as follows:

$$[RR'NCH_2CH(OH)CH_2O(CH_2)_3Si(CH_3)_2]O+[(CH_3)_2SiO]_4 \rightarrow$$
$$RR'NCH_2CH(OH)CH_2O(CH_2)_3Si(CH_3)_2]O[Si(CH_3)O]_n -$$
$$-Si(CH_3)_2(CH_2)_3OCH_2CH(OH)CH_2NRR' \tag{6}$$

Where: n=40. R=R'= C_2H_5-(XI); n=40, R= CH_2=CH-CH_2-, R' =H(XII), and n=40, R=R'= $(CH_3)_2C=N(CH_2)_2$ (XIII).

At co-oligomerization of D_4 with compounds VIII-X initiated by STMA with addition of DMFA at 80°C the conversion of D_4 reaches the maximum (94-96%) for 1.5, 3.5 and 2 hrs correspondingly. The rate of D_4 consumption till the conversion of 50-60% can be described by the first order equation and is equal to 8.5, 2.77 and 7.2 x 10^{-4}/sec correspondingly.

At oligomerization of D_4 with compounds IX and X the value of the reaction mixture η_{in} maximum is achieved after 2, and 2.5 hrs. Correspondingly and than decreases gradually to a constant level after 10-12 hrs. At co-oligomerization of D_4 with compound VIII the η_{in} reaches maximum value after 3.5 hrs and keeps the level for more than 40 hrs.

The molecular mass, hydroxyl group content and content of titrated nitrogen were determined for compounds XI-XII after removal of the low molecular components. The MM was determined by the ebullient method

Table 2. Composition and properties of co-oligomerized compounds XI-XII*

compaund	Original compaunds		[η]	MM**		η$_{in}$	N,%		-OH,%	
				Calcul	Found		Calcul	Found	Calcul	Found
XI	D$_4$	IX	0.042	3320	3400	0.043	0.84	0.84	1.02	1.10
XII	D$_4$	VIII	0.041	3288	4600	0.041	0.85	0.91	1.03	1.14
XII	D$_4$	X	0.044	3556	3700	0.045	2.36	2.38	0.96	1.00

* Co-oilgomenzation was performed in presence of initiator STMA (0.1 mass % and DMFA-1.0 mass %. ** Calculaled according to [12].

The increased MM value of compound XII could be connected with presence of oligomeric alcohols forming by the reaction of the secondary end amino groups.

The advantage of using organo-silicone compounds with amino groups on molecule ends was shown before [16]. However, when using n-phenylene diamine derivatives (n-PDA) polymerized by dihydroquinollne (DHQ) or their combinations it was found their significant volatilization and leaching front compounds. Besides, they have an ability to Initiate heat aging destruction of cured rubbers at high temperatures.

Table 3. Physical prcipelties of vulcanizates containing compounds VIII. IX.XI.and XII

Parameters	Control comp	Compounds			
		VIII	IX	XI	XII
Mod. at 300%. Elongation, MPa - at room temp.	4.4	5.1	5.0	8.5	9.0
Tensile Strenght MPa - at room temp. - heat aged*	14.6 9.5	16.3 14.4	17.0 13.9	23.0 18.4	22.9 18.3
Elorgation at break, % - at room temp. - heat aged*	660 350	690 511	695 545	550 390	545 370
Fatigue life, kcycles**	136.8	219.7	239.8	189.7	185.9

*Heat aging conditions: 100°C, 96 hrs-

** Repeated stretching at ε=100% after sample aging at 100°C for 96 hrs.

We have found that incorporation of compounds VIII, IX, XI and XII into rubber compound significantly improves tensile and fatigue properties and heat aging resistance (see Table 3).

It is seen that tensile strength of compounds modified with ammohydroxysiloxanes is much higher compared with control compound without any modifier. Especially, this is noticeable after heat aging. The tensile strength retention after heat aging for modified compounds is in a range of 80 to 88% compared with 65% for control compound. The same picture is with elongation at break. The parameter retention after heat aging for modified compounds is 68 to 78% while it is only 53% for the control compound.

Fatigue life is one of the most important properties at tire compounds. We can see from data in Table 3 a significant advantage (1.4 to 1.8 times) of the modified compounds over the control compound on the property.

We have found another application for the modifiers-coating of commercial buildings floors. The last ones undergo numerous washing with water and different cleaning products. To increase the hydrophobicity of the floors and their resistance to synthetic cleaning products (SCP), the compounds VIII, X and XIII were added to a polymer composition of the floor coating (ED-20, active dilutant - glycidyl ethers of the higher isomeric acids, keto-imine solidifier).

The coating test results are shown in Table 4. It is seen that the new compositions are chafacterized by higher hydrophobicity, lesser water absorption and enhanced resistance to action of the SCP. At the same time, the strength properties of the new composition are the same as the ones of the control composition without modifiers.

Table 4. The floor coating test results.

Property	Control coating	Modifer		
		VIII	X	XIII
Emulshion stability. Mo	32	32	32	32
Composition Life, hrs	10	10	12	10
Sodification time, hrs	24	24	24	24
Impact strength, N/m				
- original	5	5	5	5
- after immersion in SCP	0,5	5	5	5
Flex strengh, mx10^3				
- original	1	1	1	1
- after immersion in SCP	1	1	1	1
Wettability, 1/sm	4	8	10	10
Adhesion, rel.units	1	1	1	1
Wetting angle, deg.	53	78	91	93
Water absorpt, 24 hr,%	2.5	1.1	0.8	1.4
Appearence after immersion in SCP	Dull	Same	Same	Same

CONCLUSIONS

The interaction of epoxyorganosiloxanes with amines has been investigated. It was shown that the reaction ability of amino group hydrogen atoms in primary amines significantly differ and depends on the nature of organic radical at the nitrogen atom. This conclusion is based on the composition of reaction products. The yield of the products significantly depends on the ratio of original reagents.

The co-oligomerization of oktamethylcyclotetrasiloxane with aminohydroxysiloxane study showed that the reaction can be significantly promoted by addition of dimethylformamide.

The use of aminohydroxysiloxanes in rubber compounds allows to significantly increase some of very important rubber properties - tensile strength, fatigue life, modulus, heat aging resistance.

The floor coating compositions containing aminohydroxysiloxanes have enhanced hydrophobicity, decreased water absorption and improved resistance to synthetic cleaning products.

REFERENCES

1. Hedric J.L., 8. Heidar, D.C. Hofer, and J.E. McGrath, *J.Amer. Chem. Soc.*, Polym-Prepr., V.27, N.2,1986, p.203.
2. Senger J.B., 1. Jilgor, J.E. McGrath, and R.A. Patsiga, *J.Appl.Polym.Sci.*, V.38, No.2, 1989, p.373.
3. Matisons J.G., and Ar. Provatas, *Macromolecules*, V.27, 1994, p.3397.
4. Shkol'nik M.l., D.U. Murachashvili, L.M. Khananashvili, V.M. Kopylov, N.l.Tsomaia, and B. Sokol 'skaya, *Vysokomol. Soed.* A. V. 3 7, No. 4. 1995. p.571.
5. Flammarsheim H.J., H.H. Horold, K. Rellstedt, and J. Klee, *Macromol. Chem.*, V. 183, No.1, 1983, p.113.
6. Tighzert H.L., P. Bertikat, B. Chabert, and Q.T. Pham, *J. Polym.Sci.: Polym Letters Ed.*, V.20, No.8, 1982, p.417.
7. Murachashvili D.U., L.M. Khananashvili, V.M. Kopylov, N.I. Tsomaia, and M.l.Shkol'nik, lzv. AN Gruzii, *Ser. Chim.*, V.l9, No.3-4,1993, p.201(RUS).
8. Riffle J .S., I. Yilgor, C. Tran, G-L. Wilkes, and J.E. McGrath, *Amer.Chem.Soc., Symp.Ser.*.1983, p.21.
9. Bazant V., V. Chvalovsky, and J. Rathousky, *Organosilicon Compounds*, V.22, 1965, p.234, Praha.
10. Plueddemann E.P., and G. Fanger, *J.Amer.Chem. Soc.*,V.81, 1959, p.2632. Jay R.R., J.Anal.Chem, V.36. 1964. p.667.
11. Kreshkov A.P., V.A. Bork E.A. Bondarevskaya, L. V. Myshlyaeva, S. V. Syavtsillo, and V.T. Shemyatenkova, "Manuel for Analysis of Organosilica Compounds", *M. Goschimizdat*, 1962(RUS).
12. Emanuel N.M., D.G. Knoppe, *"Chemical Kinetics Course"*, M., Vysshaya Shkola, 1984(RUS).
13. KhananashviliL.M., D.U. Murachashvili, D.A. Girgvliani, Ts.N. Vardosanidze, N.I. Tsomaia, and E.G. Markarashvili, *Interns. Polym. Mater.*, V.33, 1996.p.37.
14. Murachashvili D.U., V.M. Kopylov, L.M. Khananashvili, M.I. Shkol'nik, N.I. Tsomaia, R.V. Volkova, and A. A, Savitskii, *Vysokomol.Soed.* B, V.31,1990.p.168(Rus).
15. Tati Sigemitsu, Patent Application 61-229884 (Japan), *Ref. Zhurn. Chim*, 1987, 20 y, 44n.

WATER TRANSPORT, STRUCTURE FEATURES AND MECHANICAL BEHAVIOR OF BIODEGRADABLE PHB/PVA BLENDS

A. A. Olkhov*, S. V. Vlasov*, A. L. Iordanskii*[1], G. E. Zaikov*, and V. M. M. Lobo**

* - All-union Institute of Chemical Physics (Russian Academy of Sciences) Leninskii pr.4, B334.Moscow 119991, Russia;

** - Coimbra University, Coimbra 3000. Portugal.

INTRODUCTION

In recent decades, bacterial poly(3-hydroxybutyrate) [PHB] and related bacterial poly(hydroalkanoate)'s [PHA's] have attracted much attention as biodegradable and bio-compatible materials with potential applications in packaging, agricultural, as well as medical areas [1,2]. In contrast to synthetic polymers, PHA's including PHB have the essential advantage of being renewable resources without the use of petrochemical feedstock. Moreover, this class of polymers are biodegradable i.e. PHA's can be completely digested and metabolized by a wide variety of bacteria and fungi in living body or soil and in other environments [3 - 5].

To improve the mechanical behavior of PHB and simultaneously to depress an expenditures in its production, the modification can be made through the PHB blending with other relevant polymers. Resulting polymer blends are potentially able to gain properties to be different from properties of parent blend-forming polymers. In contradistinction to traditional biodegradable polymer blends such as the starch-based plastics, PHA's posses a useful combination of transport and mechanical characteristics.

This work is devoted to the study of water diffusion in PHB - polyvynilalcohol blends and determination of relationship between transport parameters and mechanical properties of these blends. Recently authors [6] have studied PVA-PHB blends prepared by casting from the solvent dissolving both polymers simultaneously. They founded that the structures of these polymers changed and crystallinity for each poymer was decreased. It is worth to note

---

[1] To whom to send communication,the e-mail address is iordan@chph.ras.ru

that preparation of PVA-PHB blends via solution casting is poor as technological method if compare with melt extrusion. Moreover, using solvents, a serious hazard to human health and environment protection. This procedure also demands additional energetic and material expenses to regenerate the solvent used. On this bases we have studied the properties of PVA-PHB blends obtained by extrusion of the two component polymer melts.

EXPERIMENTAL

The study is concerned with PVA 8/27 Russian trade mark and PHB Biomer Krailing Germany Lot M-0997. The residual acetate group concentration and Na acetate salt concentration in PVA comprise 8.2 and 0.04 % wt. respectively. Molecular weight of PVA is 64 000g/mol. with melting point equals 146°C . The main characteristics of PHB are MW = 340 000, Tm = 178 °C (DSC data), crystallinity degree is 78% (DSC) or 69 % (X-ray technique).

Blending

The loaded concentration of PVA:PHB ingredients varied as 100:0, 90:10, 80:20, 70:30, 50:50, and 0:100. The blends were produced using a single screw extruder, ARP-20 with L/D = 25, diameter = 0.20 cm. Electricity heating was used to obtain 180° C flat extrusion profile. The components were first premixed in Brabender Plasticorder PCE330 at 170 ° C and at 60 rpm rotor speed after drying the ingredients in an air oven at 101 ° C for 8 h. The screw rotation was 100 rpm. The films obtained with final thickness 60 mkm allowed to air cool to room temperature.

Characteristics of the films are studied by DSC technique with Metler PR4000 calorimeter at heating rate 20°/min, wide angle x-ray scattering method (WAXS) at two different directions : parallel and normally to film surface. Details of X-ray measurements are described elsewhere [7]. The tensile modulus and elongation at break of the films were determined from measurements on an Instrone 1122 tensile apparatus. The drawing speed was 1.0 cm/min and the results were averaged at least five tests.

Water Vapor Permeability Experiments

Permeation of water vapor was measured at 23°C using the a regular two-compartment cell especially designated for PHB films [8]. The relative humidity in feed compartment is maintained constantly at 90%. Water content in registration compartment was very close to zero. Amounts of water transferred through polymer films are determined by weighting of KOH as absorber of water. The deviation of 5 parallel measurements for each experimental point is averaged 0.85 %. The sensing device was accurate to ±0.0001 g at 23°C .

RESULTS AND DISCUSSION

The results of DSC scans for PHB_PVA blends and the parent polymers which were prepared by extrusion at various proportion of the components are given in Table 1. The low-temperature transition between 24 and 54°C may be related to the glass transi-tion temperatures (Tg) for both parent polymers (for PHB is 24.1°C and PVA is 53.9 °C) and polymer segments of these polymers interacting in blends, see [9]. This perceptible shift of Tg can reflect the tendency for miscibility of the components. In detail, thermo physical characteristic analysis will be presented in our forthcoming paper [10].

Along with thermophysical data and the transparence in PHB-PVA films observed at 0 - 30 % concentration interval, the findings of x-ray (WAXS) method show that each of the components is capable of forming the own crystalline phase. Analysis of diffractograms (Fig.1) allows extracting in general spectra the reflexes which pertain to individual crystalline phases of PHB and PVA simultaneously. At all proportion of the components in the blends, PHB conserves the elementary cell parameters a = 0.576 nm, b = 1,32 nm, and c = 0.596 nm which correspond to orthorombic elementary cell [11]. The PVA reflexes are typical for quasi crystalline modification (the gamma-form) constructed by parallel-oriented macromolecules in dense packaging [12]. On the diffractogram in Fig.1 reflex at S = 2.21 nm^{-1} corresponds to the own phase of PVA.

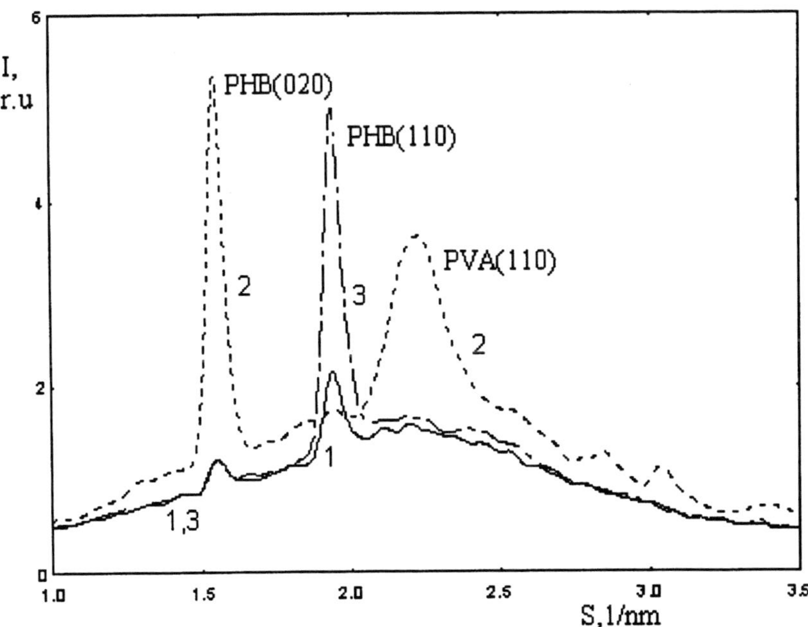

Fig.1. WAXS diffraction patterns for extruded film prepared from PHB-PVA blend at 20 % wt PHB. Measurements were provided along the extrusion direction (1), normally to the extrusion direction (2), and at 20° angle relative to extrusion direction (3). Until 30% of PHB in the blends, the X-ray pattern shows evidence of *cylindrical texture* in PHB.

Table 1. Glass transition temperatures of blends with different content of PHB.

PHB Content, %wt	0	10	20	30	50	100
T_g,° C	53.9	32.7	28.5	25.3	25.9	24.1

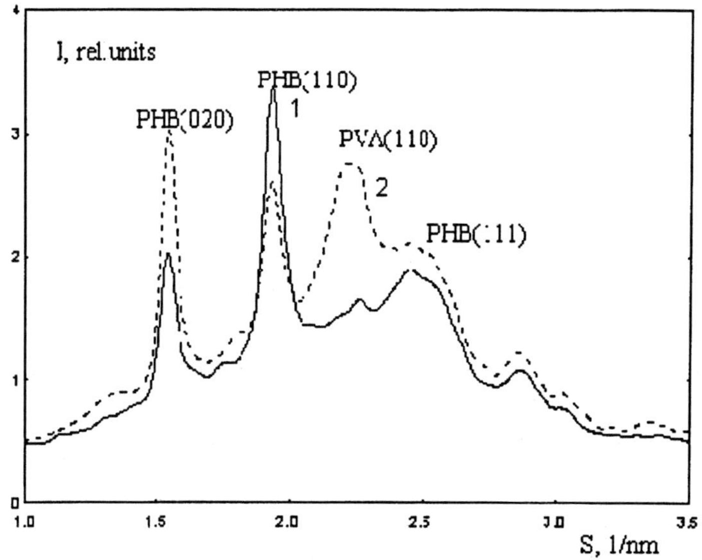

Fig.2. WAXS diffraction patterns for extruded film prepared from PHB-PVA blend at 30% wt PHB. Measurements were provided along the extrusion direction (1) and normally to the extrusion direction (2). At this concentrations the crystalline structure of PHB in blend is *isotropic*.

WAXS measurements were taken at the different orientation of film position relative to X-ray irradiation beam. For all samples the diffractograms reveal the existence of axial cylindrical texture in PVA. The axis of texture coincides with extrusion direction and hence the PVA molecules in quasicrystalline fields oriented along extrusion direction. In samples with 10 and 20 % wt of PHB a well-defined axial texture of PHB crystallites is evidence where the texture axis coincides with direction of extrusion as well. However, the PHB crystallites oriented relative to the texture axes so that the extrusion direction in line with the axes a of elementary crystalline cell. Hence, the axes of PHB molecules are normally oriented relative to extrusion direction.

Diffractograms of the samples containing 30 and 50 %wt of PHB show that the most part of crystalline phase in PHB is isotropic, without texture and only residual amount of oriented and textured crystallites is present in 30%wt- PHB-contained sample.

These findings allow concluding that in the PHB-PVA blends at 30 %wt PHB content the structural transition from textured to isotropic crystalline state occurs. Such transition could be attributed to phase inversion of polymer matrix taking place in the same concentration range, about 30 wt% of PHB. It is common knowledge that in the range of phase inversion both crystalline and physical properties of polymer blends are changed, see e.g. [13]. In this work we have studied the effect of structural inversion on mechanical behavior of the blends at different concentrations of PHB. The drastic decrease observes on the curves : 1) tensile strength - concentration and 2)elongation at break - concentration (Fig. 3, curves 1 and 2 respectively) in the same concentration interval. Besides, on the curve reflecting the dependence of elastic modulus on PHB concentration there is the minimum located in the same concentration interval near 30 % wt where the phase inversion proceeds as it shown in Fig. 4. The following increase of the blend modulus reflects the increase of concentration of structural elements (e.g. crystallites of PHB) which are responsible for high-modulus behavior of blend matrix as whole system. At the low PHB concentrations in the blends, their behavior at rupture is preferably determined by the mechanical properties of PVA while at the PHB concentration more than 30 %wt these characteristics are closely analogous to the behavior of PHB matrix. The results presented in Figures 3 and 4 do not contradict the physical concept of phase inversion involving both crystalline fields and intercrystalline (amorphous) fields in the blends.

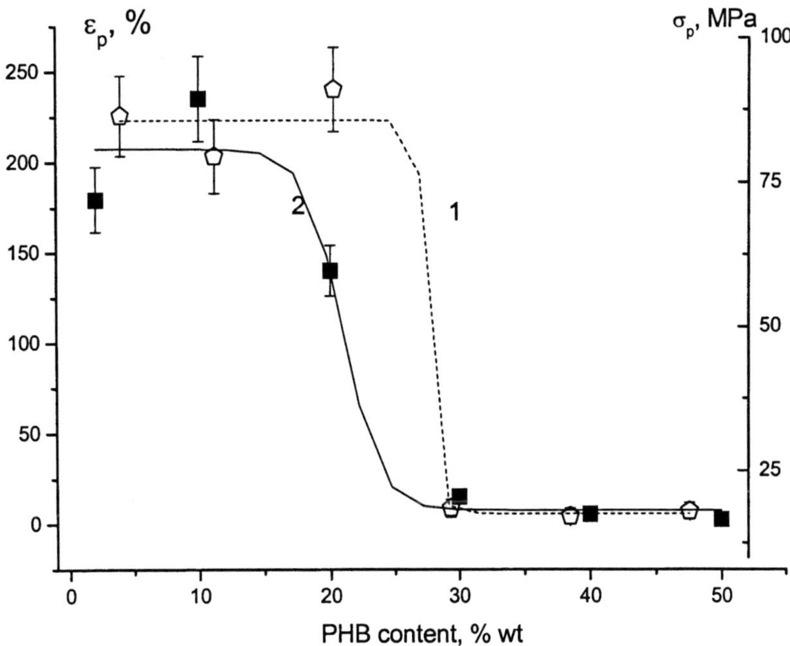

Fig.3. Mechanical characteristics of PHB-PVA blends at different concentrations of PHB. Tensile strength (1), σ_p :   ; elongation at break (2), ε_p :   .

The transformation of the amorphous fields has to effect such important characteristics as permeability and diffusivity of water. The transport processes proceed exclusively in amorphous part of any blend matrix and hence this process will be structure-sensitive relative to change of structural organization in inter crystalline fields.

Kinetic curves of vapor water permeation through the blend films at different ration PHB/PVA presented in Fig.5. Contrary to permeability through parent PVA films, all permeability curves through blends have three specific range corresponding to three different ways of water diffusion.

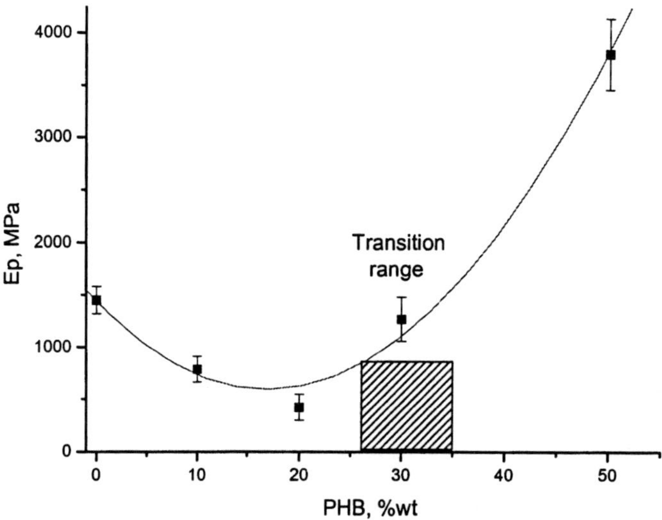

Fig.4. Effect of PHB concentration in the PHB-PVA blends on modulus of tension.

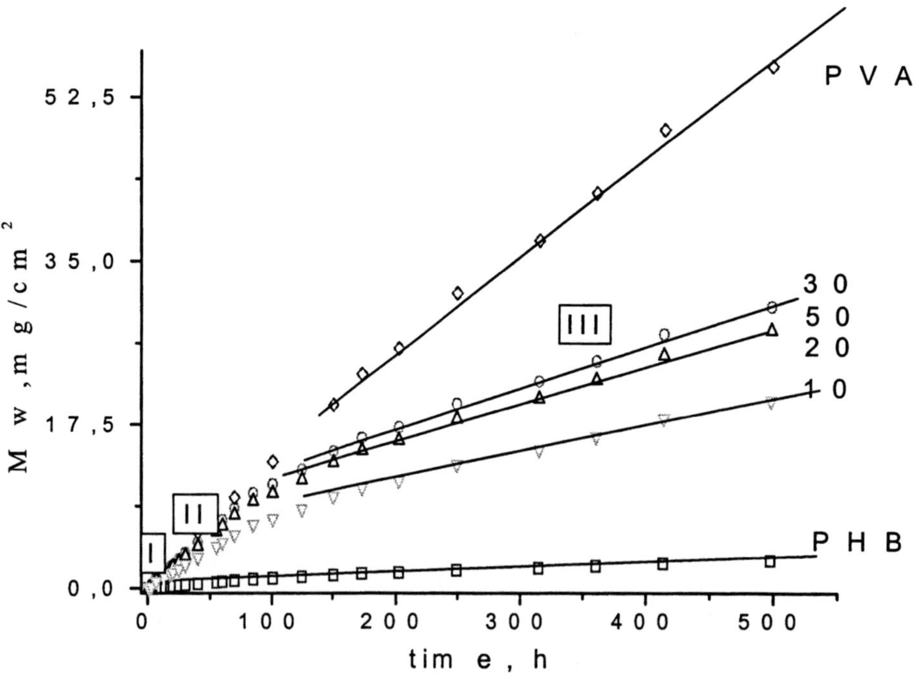

Fig.5. Kinetics curves of water permeability into films containing different ratio PHB/PVA.

The initial range (I, see Fig.5) features the low rate of water permeation where diffuison is conjugated with immobilization of water molecules on functional groups of PHB (ester groups) [14] and more intensively on hydroxyl groups of PVA [15]. In this temporal interval of time the transitional flux of water takes place that is typically for all hydrophilic polymers. The next intermediate range (II) determines the quasi steady-state regime of transport, where water diffusion is complicated by residual structural relaxation more typical for PHB molecules. A rise in vapor water permeation is dictated by both the increase of free volume ratio and segmental mobility in the blends. The last two effects result from screening the functional groups by absorbed water molecules [16] as well as redistribution of hydrogen bonds in the blends as response on water affect [17]. The rupture of hydrogen bonds formed initially between ester (PHB) and hydroxyl (PVA) or between two hydroxyl groups as the effective crosslinks promotes swelling in the blends and as consequence an increase of both water diffusivity and water equilibrium sorption. The third range of permeability curves can be recognized due to the inflection point observed for all samples containing PHB. It seems likely that to this moment the structural relaxation is completed and the water transport proceeds in accordance with regular diffusion mechanism [18].

Increasing the PVA ratio increases actually the hydroxyl group concentration in the blends. A rise of total hydrophilicity in the blend films determines a monotonic increase in equilibrium water sorption without visible point inflections or extremums, see Fig.6. In spite of possible transformation of structure on crystalline and supermolecular levels, that was revealed by WAXS and mechanical testing above, the equilibrium sorption of water molecules is sensitive to the molecular structure only. Amount of water molecules sorbed in the blends is determined by nature and concentration of the functional groups. On the other hand, with a rise of PHB content the hydroxyl concentration in polymer systems is decreased that is the replacement of the hydroxyl groups with ester groups of PHB. So the groups possessed by PVA with high affinity to water molecules are replaced with the ester groups of moderate affinity.

The permeability of water vapor as function of PHB content has maximum at about 40 %wt of PHB as it can see in Fig. 7. Probably, in this range that is very closed to 30 % range of phase inversion described above, the structure and morphology of blends have the maximal amount of defects. In the range covered 10-30 % PHB, the deviation from additive relation of water permeability is clearly observed. Even the first portion of PHB blended with PVA in ratio 1:9 depresses drastically the water permeabilty that could attest the partial interaction of the components at interval beneath 30 % of PHB. Most likely, the interaction occurs between the components in amorphous state, whereas the crystalline fractions of both components form the own fields when the PHB and PVA crystallites do not influence each other. As it follows from above results, at inversion point the PVA crystallites did not change their textured organization whereas morphology of PHB has transformed from cylindrical to isotropic state (Fig.'s 1 and 2). Structural disorder in isotropic matrix of PHB has been observed in our works recently by WAXS, FTIR spectroscopy and quartz - microbalance sorption methods [19-21]. Besides, we can not exclude the leakage of water flux through defect zones formed on the border between two components just as it happened in PHB-PELD blends described in our work [22]. The sharp build-up of heterogeneity in PHB-PVA blend films in 30 - 50 %wt interval sets one assuming an intricate mechanism of water transport including both the proper diffusion and the transport through porous areas formed of structural elements of the blended components. In this situation we has to treat these coefficients only as effective transport coefficients.

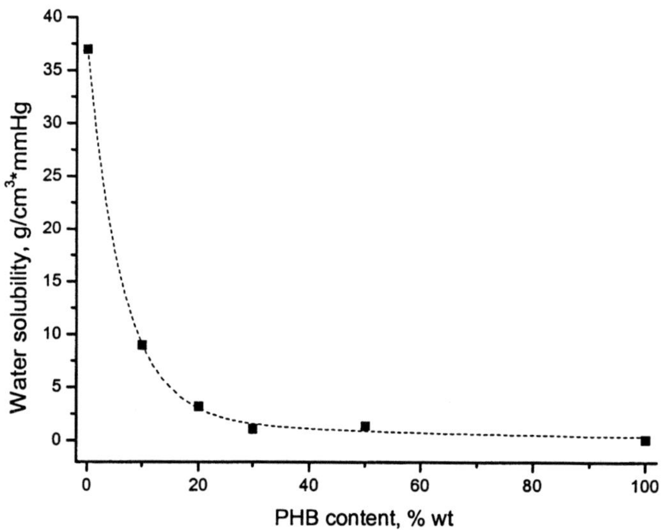

Fig.6. Water equilibrium sorption in the blends with different PHB concentration.

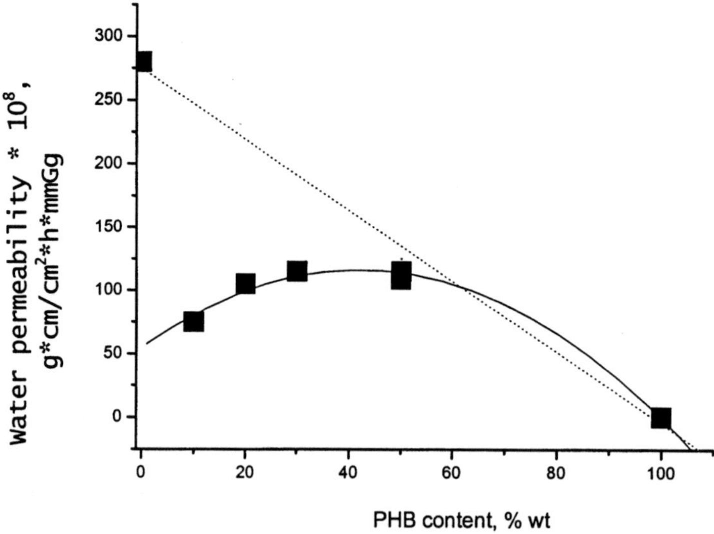

Fig.7. Water permeability through the films on the base of PHB-PVA blends at 20°C. Dashed line describes the additive relation between water permeability and PHB content.

CONCLUSIONS

A blend design on the base of polymer components with essentially different hydrophilicities is of academic and industrial interest, primarily, for the control of such critical but scantily known characteristics as water permeability, its equilibrium sorption and diffusional mobility. The higher the difference between polarity of blended macromolecules, the harder to achieve a scantily miscible blends' production, even. It is worth to point out that on retention of mechanical properties, as it has been shown for PHB-PELD reinforced blend [22], the heterogeneity or the microphase separation promote achieving the chemically corrosive degradation and biodegradation or bioerosion. It is on morphological and transport levels that the control of service properties for novel biodegradable polymer blends can be realized.

In spite of limited concentration interval of partly miscible blending (no more than 30%wt. of PHB) these PHB-PVA blends are of great interest as novel biodegradable films and coatings. Additionally, the blends based on water-soluble and biocompatible PVA and friendly environmental and biocompatible PHB could be treated as new generation of environmentally protected materials in packaging industry, agricultural application as well as biomedicine areas.

REFERENCES

1. Sudesh K., Abe H., and Doi Y. Prog. *Polym. Sci.* V.25 1503-1556 (2000). Synthesis, structure and properties of polyhydroxyalkanoates: biological polyesters.

2. Holmes P.A. In *Developments in Crystalline Polymers. II.* (Eds. D.C. Bassett) Elsevier Appl.Sci. London. 1 – 65pp. 1988. Biologically produced (R) –3-hydroxy-alkanoate polymers and copolymers.

3. Barak P., Coqnet Y., Halbach T.R., and Molina J.A.E. J. Environ. Qual. V.20, 173- 179, (1991). *Biodegradability of PHB(co-HV) and starch-incorporated polyethylene plastics, films in soils.*

4. Mergaert J., Webb A., Anderson C., Wouters A., and Swings J. Appl. Environ. Microbiol. V.93, 3233-3238, (1993). *Microbial degradation of poly(3-hydroxybutyrate) and of poly(3-hydroxybutyrate-co-3-hydroxyvaleriate) in soils.*

5. Timmins M.R., Lenz R.W., and Fuller R.C. Polymer. V.38, 551-562, (1997). *Heterogeneous kinetics of enzymatic degradation of poly(3-hydroxyalkanoates)*

6. L.Zhang, C.Xiong, and X. Deng. *J.Appl. Pol.Sci.* V.56, #1, 17-25 (1995)

7. A.V.Krivandin, O.V.Shatalova, and A.L.Iordanskii. *Polym.Sci. Ser. B.*, V.39, #3-4, 102-105 (1997). Texture of poly(3-hydroxyalkanoates) membranes.

8. P.P. Kamaev, A.L. Iordanskii, A.M. Vasserman, A.V. Krivandin. Intern. Discus. Meeting. "Advanced Methods of Polymer Characterization: New Developments and Applications Industry." Abstracts. March 15-17, 1999. Mainz. P. 3.19. *Combination of the structural and diffusional methods in system poly(3-hydroxybutyrate) - water.*

9. D.R.Paul and C.B.Bucknal. Polymer Blends. John Wiley 2000. Vol 1.

10. V.V.Yakovlev, A.A.Ol'khov, S.V.Vlasov, and A.L.Iordanskii *Macromolecular Symposia.* 2002 in Press.

11.   K.Okamura and R.H.Marchessault. *X-ray structure of poli-betta- hydroxybutyrate. In: Conformation of biopolimers. Vol.2.* NY. Academic Press. 1967. P. 709-720.

12.   A.L.Iordanskii, T.E.Rudakova, G.E.Zaikov. *Interaction of polymers with corrosive and bioactive media.* VSP. Utrecht-Tokyo. 1994. 290p.

13.   D.R.Paul and C.B.Bucknal. *Polymer Blends.* John Wiley 2000. Vol 2.

14.   A.L. Iordanskii, P.P. Kamaev, and G.E. Zaikov. *Journ. Appl. Polymer. Sci.* V.73, 981-985 (1999). *Immobilization influence on the water sorption and diffusion in poly(3-hydroxybutyrate).*

15.   C.M. Hassan and N.A.Peppas. Structure and applications of poly(vynilalcohol) hydrogels produced by conventional crosslinking or by freezing/thawing methods. In: Biopolymers. PVA Hydrogels. *Advances in Polymer Science.* Springer Verlag. Berlin-Heidelberg-NY. 2000. Vol.153. 37-66.

16.   M.Morra and C.Cassinelli Hydrophilic polysaccharides coatings. In : *Water in Biomaterials Surface Sciences* (Ed. by M.Morra). John Wiley and Sons Ltd. Chichester-NY-Wenheim-Brisbane-Singapore-Toronto. 2001. p 353-387.

17.   E.A.Vogler. Biological properties of water. . In : *Water in Biomaterials Surface Sciences* (Ed. by M.Morra). John Wiley and Sons Ltd. Chichester-NY-Wenheim-Brisbane-Singapore-Toronto. 2001. p 3-24.

18.   J.Crank and G.S.Park. *Diffusion in Polymers.* Academic Press. London-NY. 1968. 452p

19.   O.Miguel, J.J.Iruin. *J.Appl.Polym.Sci.* V.73, 455-468. (1999) Water transport properties in poly(3-hydroxybutyrate) and poly(3-hydroxybutyrate –co-3-hydroxyvaleriate) biopolymers.

20.   A.L.Iordanskii, A.V.Krivandin, O.V.Startsev, P.P.Kamaev, and U.J.Hänggi. Transport phenomena in moderately hydrophilic polymers : poly(3-hydroxybutyrate) In: *Frontiers in biomedical polymer applications.* (Ed. R.M.Ottenbrite) Technomic Publ. Lancaster-Basel. Vol.2 1999. 63-72p.

21.   P.P.Kamaev, I.I.Aliev, A.L.Iordanskii, and A.M.Wasserman. *Polymer* V.42, #2, 515-520. (2001). Molecular dynamics of the spin probes in dry and wet poly(3-hydroxybutyrate) films with different morphology.

22.   A.A.Ol'khov, A.L.Iordanskii, G.E.Zaikov, L.S.Shibryaeva, I.A.Litvinov, and S.V.Vlasov. *Intern.J.Polymeric.Mater.* V.47, 457-468, (2000)

Crystallization Behavior of Poly(3-Hydroxybutyrate) and Ethylene-Propylene Copolymer Blends

J. V. Tertyshnaya, L. S. Shibryaeva, and G. E. Zaikov

Emanuel Institute of Biochemical Physics, Russian Academy of Sciences, 4 Kosygin Str., Moscow, 119991, Russia

Abstract

Thermal properties of poly (3-hydroxybutyrate) (PHB) and ethylene-propylene copolymer (EPC) blends were studied. In order to investigate the crystallization behavior of samples the methods of differential scanning calorimetry (DSC) has been used. In the study of PHB/EPC blends by DSC, crystallization temperature of PHB in the PHB/EPC blends decreased about 4-6 ^{0}C.From the results of the Avrami analysis of PHB in the PHB/EPC blends, crystallization rate constant of PHB in the blends decreased compared to that of the pure PHB. From the above results, it is suggested that nucleation of PHB in the blends is suppressed by the addition of EPC.

INTRODUCTION

For the last two decades, a considerable amount of works has been reported on the blends polyolefin and poly(3-hydroxybutyrate) due to their academic and commercial importance. Poly(3-hydroxybutyrate) (PHB) an aliphatic polyester synthesized by bacterial fermentation, is known to degrade fully in the environment without forming any toxic products. The PHB has a good physical-mechanical properties, being close to propylene, and an effective biocompatibility with a human organism [1]. However, the high cost and rather high brittleness of PHB materials essentially limit the areas of their applications. At present, to obtain the less expensive self-destructing materials and to reduce brittle character of PHB the blends of PHB with other polymers such as starch, cellulose, polyethylene[2], polyesters are being synthesized.

There are many studies about the crystallization behavior for miscible and immiscible polymer blends. However, the crystallization behavior of PHB/EPC blends has not been reported yet. We report here on the relationship between crystallization kinetics and energy parameters for the formation of nuclei of PHB in the PHB/EPC blends.

EXPERIMENTAL PART

Materials

The polymers used in this study were obtained from commercial sources. The PHB was manufactured by Biomer Co (Germany). The chemical structure of the PHB is well established in the literature [1] and the viscosity-average molecular weight, M_η, (2.5×10^3) was determined by intrinsic viscosimetry in a chloroform solution[3].

The granulated EPC CO-059 is a product of Dutral Co (Italy) with the following mol.% content: 67.4 (ethylene) and 32.6 (propylene).

Blending

The test samples were prepared from the blends with the following ratio of PHB/EPC: 100:0, 80:20, 70:30, 50:50, wt.% respectively.

The blends were obtained by treatment for 5 min at 150 ^{0}C using a mixing lab mill VK-6 (calender diameter-80 mm, friction coefficient-1.4). The films were prepared by pressing at 190 °C in an inert atmosphere.

Test Methods

The thermal properties of all samples were analyzed by using DSC, model DSM-2M, at the scanning rate of 16 K/min. Temperature calibration was performed using indium ($Tm=156.6$ ^{0}C). Knowing specific melting heat (ΔH_m^0) for PHB (90 J/g) [4] the crystallinity was estimated with an accuracy of about 10%.

Isothermal crystallizaton experiment of the PHB in the blends was carried out on DSC. For the isothermal crystallizaton of the blends, saples were melted at 195 ^{0}C for 3 min, and then rapidly cooled to the isothermal crystallizaton temperature.

RESULTS AND DISCUSSION

Figure 1 shows endotherms of melting for the PHB/EPC blends. Each blend film shows one main endothermic peak (besides EPC) when heated from 20 to 200 ^{0}C. It is important to note that on the cooling of blends down to room temperature, only one peak is shown in the thermogram that is possessed by PHB.

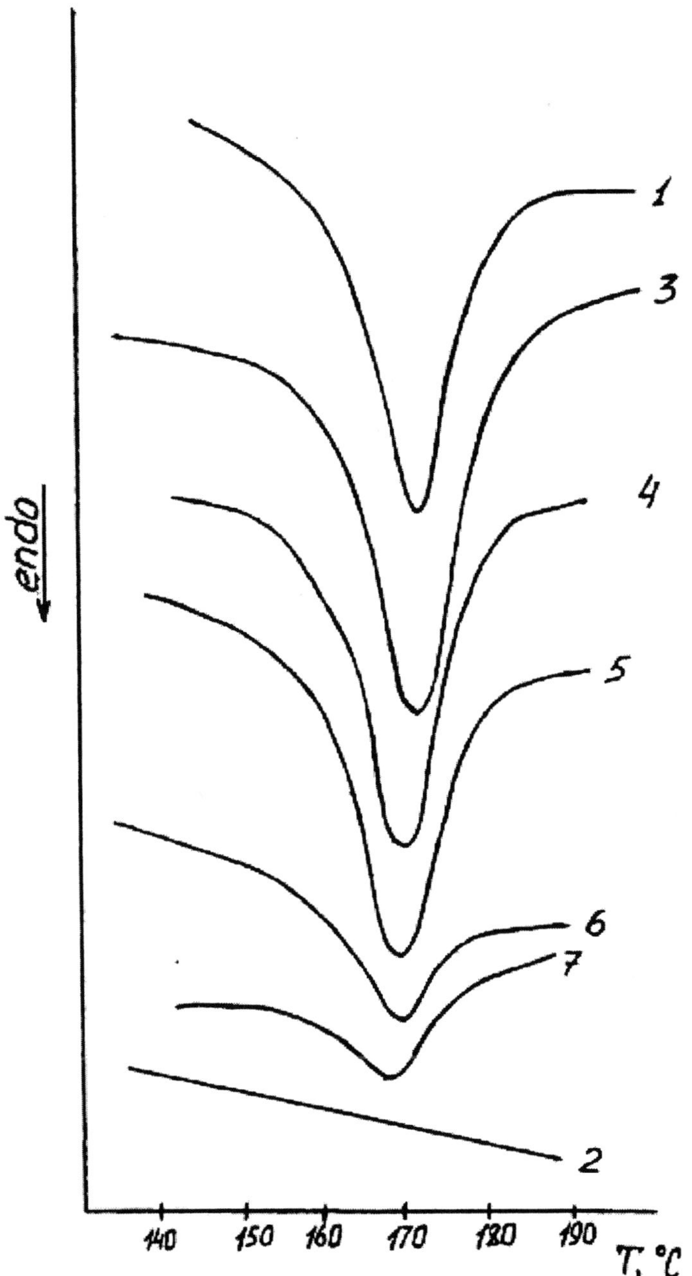

Fig.1 Endoterms of melting for PHB (1), EPC (2) and PHB:EPC blends-80:20 (3), 70:30 (4), 50:50 (5), 30:70 (6), 20:80 (7) wt.%

In the thermal analysis of PHB/EPC blends by DSC, melting temperature (T_m) of PHB in the blends is shown to be almost unchanged compared to that of pure polymer (Table 1). As for the crystallinity degree in the blends, it drops remarkably as compared to the pure polymer. From the results of T_m, ΔH_m, α %, it is suggested that the polymers have a mutual segmental solubility and this fact perhaps results in the appearance of a great interfacial layer. Also, the drop of crystallinity allows one to propose the availability of both steric hindrances occurring in melts and kinetic-relaxation problems on account of the quiet fast rate of the melt cooling in comparison with the rate of crystallization. However during our experiment by DSC the samples were heated to 195^0 (the preparation temperature was 190^0). Also the rate of cooling of the samples during the experiment was 16 ^{0}/min, it is lower than one during the preparation of films. So, it is suggested that the films can relax and the decrease of crystallinity depends strongly on the presence of the interfacial layer.

In Table 1, we can see that the T_c of PHB in the blends decreased about $2\text{-}4^0$ than the T_c of pure PHB. The decrease in T_c (PHB) of the PHB/EPC blends suggests that the EPC affects the crystallization behavior of PHB in the PHB/EPC blends.

The crystallization kinetics of the PHB/EPC blends was analyzed by the Avrami equation.

$$\alpha= 1\text{-}\exp(-kt^n),$$

where α is the crystallinity, k is the crystallization rate constant, n-the Avrami index.

Table 1. Thermal properties of PHB/EPC blended films

Blend content PHB/EPC, wt.%	$T_m, {}^0C$	ΔH_m, J/g	$T_c, {}^0C$	α_{PHB}, %
100:0	174	88,3	64	98,1
80:20	173	75,8	64	84,2
70:30	172	59,5	60	66,1
50:50	172	56,4	62	62,6

The Avrami plots of the blends at 85 ^{0}C are shown in Fig.2. From the Avrami plots of PHB in the PHB/EPC blends, the Avrami exponent *(n)* and the crystallization rate constant *(k)* were obtained and shown in Table 2.The Avrami exponent of pure PHB is about two that indicates two-dimensional lamellar growth. The *n* for all blends is the range of 2.4 to 2.6. Also, in Table 2 we can see that the *k* of pure PHB in the PHB/EPC blends has lower value than that of pure PHB. The decrease in crystallization rate constant of the PHB/EPC blends indicates that the nucleation of PHB in the blends is suppressed by the addition of EPC. Such results can be explained by negative effect of the EPC on the primary nucleation of PHB.

It is known that the presence of a second component in polymer blends has a great effect on the primary nucleation of the crystallizing component. For the immiscible blends some authors have reported that the number of heterogeneous primary nuclei of first polymer decreased with increasing concentration of second polymer in the blends (for example, for PP/LPDE blends). Chiu and authors have reported that the crystallization rate constant and the crystallization halftime of the immiscible LPDE/MFR blends are increased with the addition of MFR, probably due to the decrease of viscosity and the increase in the number of nucleation sites [5]. In our case, the deceleration of crystallization process can be explained by the presence of an interfacial layer.

From the crystallization kinetics theory:

$$G=G_0\exp(-\Delta G^*/RT-\Delta G_\eta/kT) \tag{1},$$

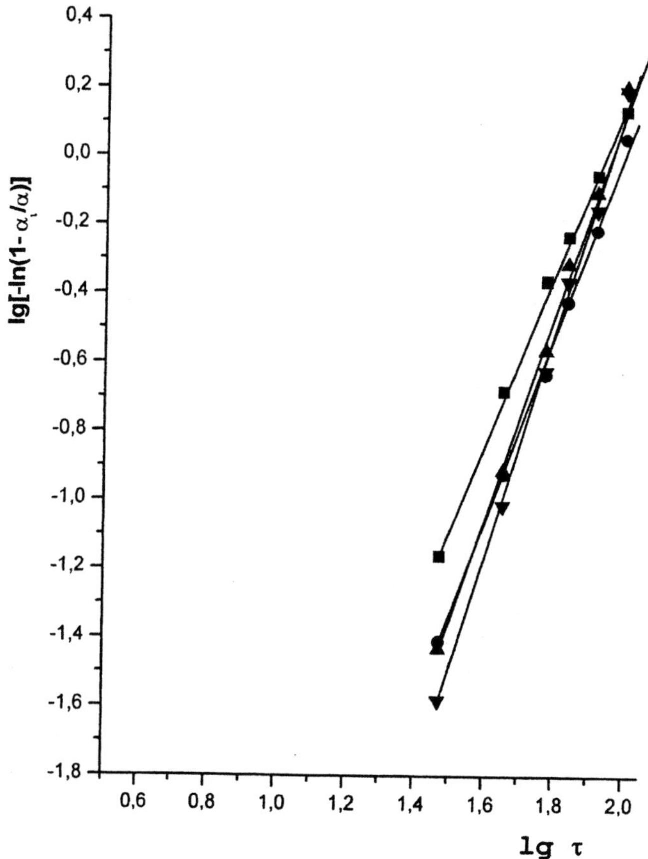

Fig.2 The Avrami analysis for PHB:EPC blends at isothermal T_c=85 ^{0}C.

where G_0 is the temperature independent constant, ΔG^* is the free energy of crystallization for forming nucleus, ΔG_η is the free energy of diffusion, R and k are the gas and Boltzmann constant respectively,

$$-\Delta G^*/RT = -WT^0_m, \quad (2) \quad \text{and} \quad -\Delta G^* = 4 \cdot b_0 \cdot \sigma_1 \cdot \sigma_2\, T^0_m/RT \cdot \Delta H_m \Delta T \rho_c \tag{3},$$

where W-energy of primary nucleation, σ_1-lateral surface energy of a nucleus, σ_2-frontal surface energy of a nucleus, and Tammann equation:

$\Delta G_\eta/RT = B/(T-T_0)$, where B and T_0 are the constants for a polymer;
$1/t_{0.5} = G_0 \cdot \exp[B/(T-T_0)] \cdot \exp[WT^0_m/T \cdot \Delta T)]$, so, using (3) and from the slopes of the plots of
$\lg(1/t_{0.5}) + [B/2.3(T-T_0)] = f[T^0_m/(T \cdot \Delta T)]$, the values of W were found and from the equation;
$W = 4 \cdot b_0 \cdot \sigma_1 \cdot \sigma_2/2.303 \cdot \Delta H_m \cdot k$, where k is the Boltzmann constant,
 the $\cdot \sigma_1$ and $\cdot \sigma_2$ were calculated as:
$\sigma_1 = 0.1 \cdot b_0 \cdot \Delta H_m \cdot$, and
$\sigma_2 = 2.303 \cdot W \cdot k/0.4 \cdot b_0^2$ where b_0-6.38$\cdot 10^{-10}$m- thickness of the polymer chain

Table 2. Crystallization halftime $(t_{0,5})$, rate constant (k) and Avrami index (n) of PHB in the PHB/EPC blends at different crystallization temperatures.

PHB:EPC, wt.%	T_c, ^{0}C	$t_{0,5}$, sec	n	k
100:0	85	70	2,34	$3,9\times10^{-5}$
	90	72	2,23	$7,6\times10^{-5}$
	92	90	2,35	$4,3\times10^{-6}$
	95	128	2,45	$2,3\times10^{-6}$
80:20	85	90	2,60	$6,3\times10^{-6}$
	90	108	2,44	$8,1\times10^{-6}$
	92	120	2,57	$3,5\times10^{-6}$
	95	135	2,65	$2,4\times10^{-6}$
70:30	85	90	2,69	$5,0\times10^{-6}$
	90	122	2,62	$3,5\times10^{-6}$
	92	145	2,64	$2,4\times10^{-6}$
	95	165	2,68	$8,9\times10^{-7}$
50:50	85	88	2,24	$6,6\times10^{-5}$
	90	135	2,54	$1,1\times10^{-6}$
	92	158	2,56	$1,8\times10^{-6}$
	95	155	2,48	$3,4\times10^{-6}$

All the calculated results are shown in the Table 3. One can see, that the values of energy parameters increase with increasing concentration EPC in the blends. Thus, from the above results, it is suggested that crystallization of PHB in the PHB/EPC blends is suppressed by presence of EPC.

Table 3. Energy parameters of the isothermal crystallization of PHB/EPC blends.

PHB:EPC, wt.%	W, K	$\sigma_1 \times 10^3$, J/m^2	$\sigma_2 \times 10^3$, J/m^2
100:0	501	6.11	97.79
80:20	497	7.03	97.01
70:30	512	7.80	99.94
50:50	564	9.03	110.10

CONCLUSIONS

In the thermal analysis of PHB/EPC blends by DSC, melting temperature of PHB in the blends is almost unchanged compared to that of pure polymer.

From the isothermal crystallization studies of PHB in the PHB/EPC blends, crystallization rate constant of PHB in the blends decreased compared to that of the pure PHB. From the measured crystallization halftime and degree of supercooling of PHB in the blends, the energy parameters of nucleation were calculated.

It was found, that the nucleation of PHB in the blends is suppressed by the addition of EPC.

REFERENCE

1. International Symposium on Bacterial Polyhydroxyalkanoates'96, Davos, Switzerland, 1996.
2. Orkhov A.A., Ivanov V.B., Vlasov S.V., lordanskii A.L., *Plast. Massy*, 6,19
3. (1998).
4. Lordanskii A.L., Kamaev P.P., and Zaikov G.E., *Int. J. Polym.Mat*er., 41,55
5. (1998).
6. Koyama N. and Dot Y., *Can. Microbiol.*, 41, 316 (1995).
7. Chiu W-Y, Wang F-T, Chen L-W, Don T-M, Lee C-Y. *Polymer Degradation and Stability*. 2000,67 p.223-231.
8. Wunderlich B. *Macromolecular physics*, vol. 2.New York: Academic Press, 1976.
9. Mandelkern L. *Crystallization of polymers*. M:Chimiya, 1966, 336 p.

AUTHOR INDEX

INDEX

non-polar unsaturated elastomers, 83
n-PDA, 86
n-phenylene diamine derivatives, 86
nucleophilic initiators, 81

O

P

Q

R